Quantum Mechanical First Principles Calculations of the Electronic and Magnetic Structure of Fe-bearing Rock-forming Silicates

Danylo Zherebetskyy

DISSERTATION.COM

Boca Raton

Quantum Mechanical First Principles Calculations of the Electronic and Magnetic Structure of Fe-bearing Rock-forming Silicates

Dissertation.com
Boca Raton, Florida
USA • 2010

ISBN-10: 1-59942-316-2
ISBN-13: 978-1-59942-316-6

Supervisors:

Prof. Dr. Michael Grodzicki

Prof. Dr. Georg Amthauer

Reviewers:

Prof. Dr. Michael Grodzicki

Prof. Dr. Werner Lottermoser

Disputants:

Prof. Dr. Werner Lottermoser

PD. Dr. Mag. Günther Redhammer

ACKNOWLEDGEMENTS

I would like to thank my supervisors Prof. Dr. Michael Grodzicki and Prof. Dr. Georg Amthauer for their guidance, support and encouragement they gave me during my three years research stage in the University of Salzburg (Austria) with all its benefits.

I am particularly indebted to Prof. Dr. Michael Grodzicki for his patience when explaining special topics in quantum physics to me, for many interesting and exciting discussions of the electronic and magnetic structure of the silicates and for many hours in reading and commenting different parts of this manuscript.

I would also like to express my appreciation to PD. Dr. Günther Redhammer and Prof. Dr. Werner Lottermoser for their help, advice and helpful comments made during my research stage.

For many helpful discussions that resulted in the ideas introduced in this thesis and the fruitful collaboration within the theoretical part of my work, it is pleasant for me to thank here Mag. Stefan Lebernegg.

Further, I am grateful to all current and former members and associates of the Department of the Material Sciences and Physics. Thanks to Gudrun Riegler and to all the technicians in the Department workgroup for their availability in various manners.

This work was founded by the Austrian Fonds zur Förderung der wissenschaftlichen Forschung (FWF) Project Number P18805_P. All calculations were carried out at the Department of Computer Sciences in Salzburg.

This work would not have been possible without the love, support and constant encouragement of my mother Kateryna Zherebetska and entire family. Special thanks to my wife Larysa.

ABSTRACT

The focus of this thesis is the study of the electronic and magnetic structure of three representative members of Fe-bearing rock-forming silicates, viz. orthoferrosilite ($Fe^{2+}_2Si_2O_6$), almandine ($Fe^{2+}_3Al_2(SiO_4)_3$) and andradite ($Ca_3Fe^{3+}_2(SiO_4)_3$). These minerals have attracted significant attention due to their abundance in the Earth's crust and mantle, and because crystallised silicates are main components of cosmic dust which is the most abundant raw material in the Universe. For this purpose quantum mechanical first principles electronic structure calculations are performed by the most efficient DFT method in the local spin-density approximation for calculating spectroscopic data: the spin-polarized self consistent charge $X\alpha$ method.

The specific feature and strength of these investigations consist in the theoretical characterization of these complex systems based on experimental results. This means that, on one hand, experimental spectroscopic and crystallographic data are being used to judge the reliability of the calculations, whereas, on the other hand, experimental data are interpreted and explained by the theoretical results.

This work comprises seven chapters. After a brief introduction (Chapter 1) Chapter 2 describes the theoretical bases, ideas, approximations and advantages of the SCC- $X\alpha$ method and basics of the art of cluster construction. Chapter 3 considers physical bases of crystal field theory, absorption, Mössbauer spectroscopy and magnetic interactions, as well as the calculation of spectroscopic data within the frame of the SCC-$X\alpha$ method. In addition, tetragonally, trigonally and angularly distorted octahedral sites with various degrees of the distortions are calculated and analyzed. The electronic and magnetic structures of orthoferrosilite, almandine and andradite are described in the following chapters. In the case of orthoferrosilite the magnetic interactions between the iron spins within the ribbons and between neighboring ribbons are characterized. Two identical interpenetrating magnetic sublattices of circles of 10 edge-shared dodecahedra are revealed and characterized in almandine. The calculated spin structure explains and solves the controversy in the interpretation of the Mössbauer spectra of almandine below the Néel temperature. For andradite a model of the magnetic structure is proposed based on geometrical considerations and the calculated spin coupling constants for the various interaction pathways. According to this model, the magnetic structure of andradite consists of two frustrated equivalent magnetic sublattices. The spins of the Fe ions within each sublattice are coupled antiferromagnetically. The derived spin pattern explains two sextets in the Mössbauer spectra of andradite below the Néel temperature. Finally, the main results are summarized in Chapter 7.

CONTENTS

LIST OF FIGURES

7

LIST OF TABLES

Chapter 1

INTRODUCTION

Silicates are the main constituents of the Earth's crust and mantle comprising 95% of the crust and 97% of the mantle, by most estimates, and the most important rock-forming minerals. Correspondingly, about one fourth of the known minerals of the crust are silicates, and the nine most abundant elements in the Earth's crust, viz. O (61.2 %), Si (20.8 %), Al (6.4 %), Na (2.6 %), Fe (1.9 %), Ca (1.9 %), Mg (1.8 %), K (1.4 %), H (1.0 %), appear as the most major constituting atoms of the various silicate minerals. Moreover, the European Space Agency's Infrared Space Observatory discovered in 1998 that crystallised silicates are main components of dust which is the most abundant raw material in the Universe (planets, comets etc. are all made from reprocessed dust). This finding named "crystalline revolution" that is one of the main results of ESA's infrared space telescope, opens a totally new field astro-mineralogy.

Among the most abundant elements listed above only iron can carry a magnetic moment because of its partly-filled d shell, while all the other ions have a closed shell configuration in their common oxidation state. In fact, the reason for the magnetic properties of Fe-bearing silicates is that iron in both of its two common oxidation states occurs in the high-spin state with 5 unpaired $3d$ electrons (Fe^{3+} with spin S=5/2) and 4 unpaired $3d$ electrons (Fe^{2+} with spin S=2), respectively. Since the abundance of any other magnetic ion is more than an order of magnitude smaller than that of iron, e.g. Mn with 0.05 % as the second most abundant open shell ion, the discussion of the magnetic properties of minerals must necessarily focus on iron oxides and iron silicates and a proper understanding of the physics and chemistry of our planet and Universe requires information on their properties.

Whereas the magnetic properties of most iron oxide minerals, some of them known since ancient times, have extensively been investigated and are thus rather well known, the magnetic behaviour of iron bearing silicates is far less well understood. The obvious reason is the complexity of their structure so that

(i) the interpretation of the experimental results is more difficult and

(ii) reliable theoretical calculations can hardly be found.

Moreover, none of the most abundant rock forming iron silicates is magnetically ordered at ambient temperature, but almost all of them exhibit complex low-temperature magnetic structure, usually characterized by antiferromagnetic ordering with Néel temperatures T_N in the range between 1 and 100 K.

Nevertheless, the occurrence of magnetic ordering at low temperatures can still have a significant influence on the thermodynamic properties at elevated temperatures since crystal-field effects and magnetic ordering can contribute up to 10% and more to the specific heat, and thus to the enthalpy and entropy at room temperature. In general, the Mg- and Al-silicates show considerably different physico-chemical properties in comparison with the isotypic Fe-silicates, e.g. enstatite, $Mg_2Si_2O_6$, vs. ferrosilite, $Fe^{2+}_2Si_2O_6$, pyrope, $Mg_3Al_2(SiO_4)_3$, vs. almandine, $Fe^{2+}_3Al_2(SiO_4)_3$, and grossular, $Ca_3Al_2(SiO_4)_3$, vs. andradite, $Ca_3Fe^{3+}_2(SiO_4)_3$. Thus, the presence of magnetic ions in a solid solution may change its macroscopic properties in a fundamental manner. Additionally, magnetic ions provide a sensitive probe of the local chemical environment, and chemical disorder may induce spin glass behaviour.

The basic aim of the present study is to understand the origin and the details of the low-temperature magnetic structure of representative iron-bearing silicates as an important class of rock-forming minerals. Since a qualitatively correct model of the electronic structure is the necessary basis for understanding the magnetic properties, this aim is expected to be achieved by systematic theoretical investigations of the electronic structure from which spin density distributions, magnetic moments, and the parent magnetic interaction constants can be derived. These results are utilized to trace back the reasons for the different strengths of the magnetic interactions, as expected, e.g., in the different ordering temperatures T_N, to the electronic and geometrical structure. The specific feature and strength of these investigations consist in the theoretical characterization of these complex systems based on experimental results. This means that, on one hand, experimental spectroscopic and crystallographic data are being used to judge the reliability of the calculations, whereas, on the other hand, experimental data are interpreted and explained by the theoretical results. For, in many cases different physical effects contribute to a certain observable and the influence and significance of the various contributions can usually not be estimated from experiments alone, while calculations may help to settle this type of questions.

The innovative aspects of these investigations consist in quantum mechanical first principles calculations, i.e. without using empirically derived parameters. To the best of our knowledge calculations on this level of sophistication have not yet been applied for the determination of magnetic interaction constants in Fe-bearing silicates. The theoretical derivation of quantitatively correct values for the Heisenberg coupling constants for systems of such a complexity as provided by the iron-silicates is certainly the most important and challenging step towards describing and explaining the magnetic properties not just of the Fe-bearing silicates but also of other similarly complex systems. Consequently, investigating and understanding the magnetic properties and magnetic structures of iron silicates is of basic importance and will substantially contribute to the understanding of these phenomena as well as to the thermodynamic properties of Fe-bearing rock-forming silicates.

In summary, Fe-bearing silicates exhibit a variety of interesting magnetic properties reflecting peculiarities of the silicate structure. Very often minor changes of the crystal structures and chemical compositions have a great influence on the magnetic properties such as magnetic ordering temperatures, sizes of the local magnetic fields, magnetic coupling, etc. However, the complete relations between crystal structures and magnetic properties of the various Fe-bearing silicates are not yet understood.

The selection of suitable Fe-silicates for investigating magnetic ordering is dictated by several boundary conditions that must be fulfilled for successfully carrying out of this work, namely:

–   Spectroscopic data, e.g. from Mössbauer, and/or optical spectroscopy, and/or magnetic susceptibility should be available for the selected systems in order to have an independent measure for judging the reliability of the calculations.

–   The systems should be representative and relevant from the mineralogical and geochemical point of view. In this respect pyroxenes and garnets belong to the most important Fe-bearing rock-forming silicates

Based on these considerations, the orthopyroxene solid solution series, as well as end-member garnets almandine and andradite have been selected. The quantum mechanical first principles electronic structure calculations are performed by the most efficient DFT method in the local spin-density approximation for calculating spectroscopic data: the spin-polarized self consistent charge $X\alpha$ method.

In the first part of this thesis, the theoretical bases, ideas, approximations and advantages of the SCC- $X\alpha$ method and basics of the art of cluster construction are described (Chapter 2) and, furthermore, we considered the physical bases of absorption and Mössbauer spectroscopy, crystal field theory, evaluation of the main spectroscopic values within the frames of the SCC- $X\alpha$ method and magnetic interaction between atoms (Chapter 3). In addition, the splitting patterns of the $3d$ orbitals of the Fe^{2+} ion in tetragonally, trigonally and angularly distorted octahedral site for variable degrees of the distortions are calculated and analyzed. It is shown, that the point charge model for description of the crystal-field splitting is inappropriate and the model based on the overlap integrals is very appropriate for the description of the crystal-field splitting.

In the second part of this work (Chapter 4-6) the electronic and magnetic structures of orthoferrosilite $Fe^{2+}_2Si_2O_6$, almandine $Fe^{2+}_3Al_2(SiO_4)_3$ and andradite $Ca_3Fe^{3+}_2(SiO_4)_3$ are investigated and characterized by electronic structure calculations in the local spin density approximation.

Chapter 2

THEORETICAL BASES

2.1 Born-Oppenheimer Approximation

The quantum mechanical description of an electronic system requires the solution of the (time-independent) Schrödinger equation (SE). Neglecting magnetic interactions and relativistic effects, the full *nonrelativistic Hamiltonian* for a molecule, cluster or solid comprises contributions from all the nuclei and the electrons of the system:

$$\hat{H} = \hat{T}_N + \hat{T}_e + \hat{U}_{Ne} + \hat{U}_{ee} + \hat{U}_{NN}, \tag{2.1}$$

where $\hat{T}_N$ and $\hat{T}_e$ are the operators of the kinetic energy of the nuclei and electrons, respectively, $\hat{U}_{Ne}$ is the operator of the attractive electrostatic interaction between electrons and nuclei, $\hat{U}_{ee}$ and $\hat{U}_{NN}$ are the operators of the electrostatic electron-electron and nuclear-nuclear repulsion, respectively. For a system of N nuclei and K electrons, this Hamiltonian explicitly can be written as:

$$\hat{H} = -\frac{\hbar^2}{2M_I} \sum_{I=1}^{N} \nabla_I^2 - \frac{\hbar^2}{2m_e} \sum_{i=1}^{K} \nabla_i^2 - \sum_{I=1}^{N} \sum_{i=1}^{K} \frac{Z_I e^2}{|\vec{R_I} - \vec{r_i}|} +$$

$$+ \sum_{i=1}^{K} \sum_{j<i}^{K} \frac{e^2}{|\vec{r_i} - \vec{r_j}|} + \sum_{I=1}^{N} \sum_{J<I}^{N} \frac{Z_I Z_J e^2}{|\vec{R_I} - \vec{R_J}|}, \tag{2.2}$$

where M_I is the mass of Ith nucleus and m_e is the electron mass, e is the elementary charge, $\vec{r_i}$ is the position of the ith electron, $\vec{R_I}$ and Z_I are the position and the atomic number of the Ith nucleus, respectively. In equations (2.1-2.2), relativistic effects, such as spin-orbit or spin-spin coupling are neglected.

The vast majority of electronic structure calculations on molecules, clusters and solids is based on the experience that the motions of the nuclei and electrons are, in general, only weakly coupled due to the large difference in the masses of the electrons and protons (nuclei)

$$\frac{m_{el}}{M_{prot}} = \frac{1}{1836.15} \implies \frac{m_{el}}{M_{nuc}} \approx 2 \cdot 10^{-4} \dots 10^{-6}. \tag{2.3}$$

In addition, since the absolute values of the electric charge of the electron and the proton are identical, attractive and repulsive electrostatic interactions between the particles are of the same order of magnitude. Due to this difference in masses the electronic and nuclear motions take place on different time scales, i.e.:

1) the electrons move much faster, and adjust to each change of nuclear positions almost instantaneously;

2) with respect to the motion of the electrons, the ions can be assumed as frozen.

The formal separation of electronic and nuclear motion is known as *the Born-Oppenheimer approximation*, adiabatic or clamped-nuclei approximation (Born and Oppenheimer 1927). In this approximation the total wavefunction is written as the product of a nuclear and an electronic part

$$\Psi(\vec{R}, \vec{r}) = \psi_{nuc}(\vec{R})\psi_{el}(\vec{r}), \tag{2.4}$$

where $\vec{R}$ are all nuclear coordinates and $\vec{r}$ are the coordinates of all electrons. Substituting this wavefunction into the Hamiltonian (2.1) yields two separate SE, viz.:

$$\left(\hat{T}_e + \hat{U}_{ee} + \hat{U}_{Ne}\right)\psi_{el} = E_{el} \cdot \psi_{el} \tag{2.5}$$

which refers to the electronic system, where the energy E_{el} and wavefunction ψ_{el} depend parametrically on the nuclear positions $\vec{R}$, and

$$\left(\hat{T}_N + \hat{U}_{NN} + E_{el}\right)\psi_{nuc} = E \cdot \psi_{nuc} \tag{2.6}$$

describing the nuclear motions.

The corresponding electronic Hamiltonian in the Born-Oppenheimer (BO) approximation is

$$\hat{H}_{el} = -\frac{\hbar^2}{2m_e}\sum_{i=1}^{K}\nabla_i^2 - \sum_{I=1}^{N}\sum_{i=1}^{K}\frac{z_I e^2}{|\vec{R}_I - \vec{r}_i|} + \sum_{i=1}^{K}\sum_{j<i}^{K}\frac{e^2}{|\vec{r}_i - \vec{r}_j|}. \tag{2.7}$$

Solving SE (2.5) for the electronic wavefunction gives the eigenvalue $E_{el}(\vec{R})$ of the electronic Hamiltonian (2.7). One must remember during the treatment of the electronic SE that the nuclei move relatively slowly and each new relative positions of the nuclei $\vec{R}$ provide a new solution to the SE. Accordingly, the total energy of the system E_T is the sum of electronic energy E_{el} and the nuclear repulsion term U_{NN}:

$$E_T(\vec{R}) = U_{NN} + E_{el}(\vec{R}). \tag{2.8}$$

In this picture the electrons move in the field of the fixed nuclei (2.5) while the nuclei move on a potential energy surface (2.8), where the electronic energy compensates the nuclear-nuclear repulsion. Consequently, there is no stable nuclear system without electrons. Furthermore, solution of the SE (2.6) for the nuclear wave function leads to energy levels for molecular vibrations and rotations.

The BO approximation is the first step towards simplifying the solution of the SE. However, the approximation contradicts with respect to the nuclei one of the fundamental principles of quantum mechanics, namely the Heisenberg uncertainty principle. Since the uncertainty principle does not allow simultaneously certain impulse and certain position any particle, it is wrong to say that the nuclei are frozen having certain coordinates. Therefore, the Born-Oppenheimer approximation cannot be realized within the frames of QM. The conceptual basis of the BO approximation consists in the classical description of the nuclear motion. Correspondingly, the nuclear positions $\vec{R_I}$ are treated as parameters and not as dynamical variables. Consequently, the conceptual basis of molecular and solid state physics (structure, bond, potential curve, electronic state, molecular vibration, etc.) is not defined within the frame of a purely quantum mechanical description. Therefore, the two fundamental concepts of any theory of condensed matter, viz. structure and dynamics, are not defined within the frames of a purely quantum mechanical description (Grodzicki 2007).

2.2 Density Functional Theory

Density functional theory (DFT) is one of the most popular and successful quantum mechanical approaches to simulations of matter. In recent years, it has widely been applied for simulation studies in order to understand the behavior and properties of molecules and solids. The development of the DFT leads to important breakthroughs in chemical calculations and is a very significant contribution to the science of many-particle quantum systems, including problems of electronic structure of molecules and of condensed matter. For the pioneering contributions in developing methods that can be used for theoretical studies of the properties of molecules and the chemical processes in which they are involved, Walter Kohn who played the leading role in the development of the DFT was awarded, jointly with John Pople, the Nobel Prize in chemistry in 1998 (Kohn 1995, 1999).

The knowledge of the energies and wavefunctions of electrons in a system under study enables the calculation of optical, magnetic, thermal properties of the system, and therefore understanding the behaviour of atoms, molecules and solids. However, the wavefunction is a very complicated quantity for an N-electron system that depends on $3N$ spatial and N spin variables. This huge number of variables makes Hartree-Fock computations very difficult or impossible at all. Within conventional quantum-chemical approaches (HF and beyond), it was common wisdom that the exact description of the electronic structure requires both the *single-particle density* $\rho(\vec{r})$ and the *two-particle density* $\pi(\vec{x}_1, \vec{x}_2)$:

$$\rho(\vec{r}) = N \int \dots \int |\Psi(\vec{x}_1, \vec{x}_2, \dots, \vec{x}_N)|^2 ds_1 d\vec{x}_2 \dots d\vec{x}_N, \qquad (2.9)$$

$$\pi(\vec{x}_1, \vec{x}_2) = N(N-1) \int \dots \int |\Psi(\vec{x}_1, \vec{x}_2, \dots, \vec{x}_N)|^2 d\vec{x}_3 \dots d\vec{x}_N. \qquad (2.10)$$

In equation (2.9), $\vec{x}_i = (\vec{r}_i, s_i)$ and $\rho(\vec{r})$ is the probability to find any of N electrons within the volume $d\vec{r}_1$ (at position $\vec{r}_1$) with arbitrary spin while the other electrons of the system have arbitrary positions. In turn, the two-particle density $\pi(\vec{x}_1, \vec{x}_2)$ defines the probability to find two electrons with spins s_1 and s_2 within two volume elements $d\vec{r}_1$ and $d\vec{r}_2$ (at positions $\vec{r}_1$ and $\vec{r}_2$), respectively.

The probability density $\rho(\vec{r})$ is strictly the electron number density and not electron density, which is defined as $-e\rho(\vec{r})$. However, numerous authors speak about probability density and electron density as if it were the same thing. Such sloppy usage is stipulated by the fact that probability density is related to a random variable and the authors want to indicate that the electrons are the particles under consideration. It is clear that the integral of the probability density is the total number of electrons

$$\int \rho(\vec{r})d\vec{r}_1 = N. \tag{2.11}$$

An important property of the probability density is the observability, i.e. it can be measured experimentally by X-ray diffraction.

In turn, the pair density $\pi(\vec{x}_1, \vec{x}_2)$ contains important information about electron correlation. The integral of the pair density is the total number of non-distinct electron pairs

$$\int \pi(\vec{x}_1, \vec{x}_2)d\vec{x}_1 d\vec{x}_2 = N(N-1). \tag{2.12}$$

Since electrons are indistinguishable particles, the pair density may be written as

$$\pi(\vec{x}_1, \vec{x}_2) = \frac{N-1}{N}\rho(\vec{x}_1)\rho(\vec{x}_2)(1 - g(\vec{x}_1, \vec{x}_2)), \tag{2.13}$$

where $g(\vec{x}_1, \vec{x}_2)$ is the pair-correlation function.

With regard to this background it was surprising and of central importance that an exact representation of the quantum-mechanical many-particle problem can be constructed solely on the basis of the single-particle density (Hohenberg and Kohn, 1964).

The attempt to use the probability density instead of the wavefunction for molecular systems is almost as old as quantum mechanics (Thomas 1927 and Fermi 1927). In the first attempt within the frame of DFT, the quantum nature of electrons is taken into account only in kinetic energy while nuclear-electron and electron-electron interactions are described in a classical way by Coulomb electrostatic interaction. The fermion nature of electrons is not taken into account. The Thomas-Fermi model uses a very simple functional expression for kinetic energy which is based on the uniform electron gas. In this model, the total ground-state energy of an atom is a functional of the density

$$E_{TF}[\rho(\vec{r})] = T[\rho(\vec{r})] + U_{ext}[\rho(\vec{r})] + U_{ee}[\rho(\vec{r})] =$$
$$= \frac{3}{10}(3\pi^2)^{2/3}\int(e\rho)^{5/3}(\vec{r})d\vec{r} - Ze^2\int\frac{\rho(\vec{r})}{\vec{r}}d\vec{r} + \frac{e^2}{2}\iint\frac{\rho(\vec{r}_1)\rho(\vec{r}_2)}{|\vec{r}_1-\vec{r}_2|}d\vec{r}_1 d\vec{r}_2, \tag{2.14}$$

where $T[\rho(\vec{r})]$ is the kinetic energy, $U_{ext}[\rho(\vec{r})]$ is the external potential and $U_{ee}[\rho(\vec{r})]$ is the energy of the electron-electron repulsion. The model gives good qualitative picture of electronic structure but not the shell structure and single-particle density does not behave exponentially and it predicts all molecules to be unstable toward dissociation into their atoms.

The idea to consider the single-particle density instead of the wavefunction plays a central role in the *Density functional theory*. The next stage for this approach and the whole theory is based on the three remarkable theorems of Pierre Hohenberg and Walter Kohn (Hohenberg and Kohn, 1964).

(HK1) *The local external potential* $U_{ext}(\vec{r})$ *is a unique functional of the electron density* $\rho(\vec{r})$, *apart from a trivial additive constant,* which has modified formulation: *the energy* E_0 *of the ground state is unique functional of the ground state electron density* $\rho(\vec{r})$ (Grodzicki 1999).

In other words, if the electron density $\rho_0(\vec{r})$ of the ground state of a quantum-mechanical many-particle system is known any ground state property can be calculated (Grodzicki 1999). Additionally, it should be mentioned that the energy computed as the expectation value of the Hamiltonian $\hat{H}$ from any guessed wavefunction, ψ_t, is always higher or equal than the energy of the ground state

$$\frac{\int \psi_t^* \hat{H} \psi_t dx}{\int \psi_t^* \psi_t dx} = E_t \geq E_0 = \frac{\int \psi_0^* \hat{H} \psi_0 dx}{\int \psi_0^* \psi_0 dx}. \qquad (2.15)$$

Hence, we have to minimize the functional, $E_t[\psi_t]$, by searching through all acceptable wavefunctions, in order to find the ground state energy and wavefunction. This is a so-called *variational principle*. The proof of the Hohenberg-Kohn theorem is based on a *reductio ad absurdum* argument using this principle.

The DFT analogue of the variational principle is HK2 (second Hohenberg-Kohn theorem), which states that:

$E[\rho(\vec{r})]$ *assumes its minimum value for the correct ground state density* $\rho_0(\vec{r})$, *if the admissible functions* $\rho(\vec{r})$ *are restricted by the condition*

$$N[\rho(\vec{r})] = \int \rho(\vec{r}) d\vec{r} = N. \qquad (2.16)$$

If we define a trial one-electron density $\rho'(\vec{r})$ satisfying conditions $\rho'(\vec{r}) \geq 0$ for all $\vec{r}$ and $\int \rho'(\vec{r}) d\vec{r} = N$ than the HK2 can be formulated as:

(HK2) *The energy calculated from a trial one-electron density* $\rho'(\vec{r})$, *such that* $\rho'(\vec{r}) \geq 0$ *(for all r) and* $\int \rho'(\vec{r}) d\vec{r} = N$, *is always larger than the exact ground-state energy* E_0

$$E[\rho'(\vec{r})] \geq E_0[\rho_0(\vec{r})], \qquad (2.17)$$

and sign of equality is valid exactly when $\rho'(\vec{r})$ *is equal to the exact ground-state one-particle density* $\rho_0(\vec{r})$.

In other words, the minimum value of the energy corresponds to the exact ground-state electron density and any approximate density gives higher value of energy.

Assuming differentiability of $E[\rho(\vec{r})]$, the variation of the energy functional gives the Euler-Lagrange equation

$$\frac{\delta E[\rho(\vec{r})]}{\delta \rho(\vec{r})} = \mu, \qquad (2.18)$$

where the Lagrange parameter μ can be interpreted as the chemical potential of the system.

Both theorems are *existence theorems,* which state that the external potential and hence all properties of the many-particle system, are determined by the ground-state electron density distribution. Therefore, they do not provide any concrete form of the functional.

In the next step, a *universal functional,* which describes the kinetic energy of the electrons and interelectronic interaction, is defined as

$$F[\rho(\vec{r})] = \int \Psi^*(\hat{T} + \hat{U}_{ee})\Psi d\vec{r}. \qquad (2.19)$$

The universal functional is valid for any number of particles and independent of the external potential. With the help of the universal functional, the energy functional for a given external potential $U_{ext}(\vec{r})$ is defined as

$$E[\rho(\vec{r})] = \int U_{ext}(\vec{r})\rho(\vec{r})d\vec{r} + F[\rho(\vec{r})]. \qquad (2.20)$$

A first step to the solution of the concrete functional problem is the third Hohenberg-Kohn theorem:

(HK3) *The total energy of a N-electron system in a local, external potential $U_{ext}(\vec{r})$ can be presented in the form*

$$E[\rho(\vec{r})] = \int \rho(\vec{r})U_{ext}(\vec{r})d\vec{r} + T[\rho(\vec{r})] + U_{ee}[\rho(\vec{r})] = U_{ext}[\rho(\vec{r})] + F_{HK}[\rho(\vec{r})], \quad (2.21)$$

where $F_{HK}[\rho(\vec{r})]$ does not depend on $U_{ext}(\vec{r})$, and therefore is a universal functional composed of the kinetic energy $T[\rho(\vec{r})]$ of the electrons and the electron-electron interaction $U_{ee}[\rho(\vec{r})]$. The universal functional $F_{HK}[\rho(\vec{r})]$ has the same form as functional of ρ in all systems, so that, the N-electron problem can be split into a general, system-independent and a system-specific parts.

This problem was solved (Kohn and Sham, 1965) by the suggestion that the universal functional should be

$$F[\rho(\vec{r})] = T_s[\rho(\vec{r})] + J[\rho(\vec{r})] + E_{XC}[\rho(\vec{r})], \qquad (2.22)$$

where

$$T_s[\rho(\vec{r})] = \Sigma_i^N \int \psi_i(\vec{r}_i)(-\frac{\hbar^2}{2m_e}\nabla^2)\psi_i(\vec{r}_i)d\vec{r}_i \qquad (2.23)$$

is the kinetic energy of noninteracting electrons,

$$J[\rho(\vec{r})] = \frac{1}{2}\iint \frac{e^2\rho(\vec{r}_i)\rho(\vec{r}_j)d\vec{r}_i d\vec{r}_j}{|\vec{r}_i - \vec{r}_j|} \qquad (2.24)$$

is the classical electrostatic energy arising from the interaction between two charge densities, and

$$E_{XC}[\rho(\vec{r})] \equiv (T[\rho(\vec{r})] - T_s[\rho(\vec{r})]) + (U_{ee}[\rho(\vec{r})] - J[\rho(\vec{r})]) \qquad (2.25)$$

is the so-called *exchange-correlation energy*. In equation (2.25), T is the exact kinetic energy of the interacting system, T_S is the kinetic energy of the non-interacting system and U_{ee} is the full electron-electron repulsion energy. Hence, the exchange-correlation energy is the part containing all unknown interactions. In addition, it is possible to split the exchange-correlation energy into an exchange and a correlation contribution

$$E_{XC}[\rho(\vec{r})] = E_X[\rho(\vec{r})] + E_C[\rho(\vec{r})] \qquad (2.26)$$

It should be emphasized that usually the exchange contribution is about an order of magnitude larger than the corresponding correlation energy.

Using equations (2.22)-(2.25), the full expression for the total electronic energy within the Kohn-Sham scheme can be written as

$$E[\rho(\vec{r})] = \sum_i^N \int \psi_i(\vec{r}_i)\left(-\frac{\hbar^2}{2m_e}\nabla^2\right)\psi_i(\vec{r}_i)d\vec{r}_i + \frac{e^2}{2}\iint \frac{\rho(\vec{r}_i)\rho(\vec{r}_j)d\vec{r}_i d\vec{r}_j}{|\vec{r}_i - \vec{r}_j|} + E_{XC}[\rho(\vec{r})] + U_{ext}[\rho(\vec{r})].$$

$$(2.27)$$

The corresponding one-electron equations are denoted as *Kohn-Sham equations*

$$\left[-\frac{\hbar^2}{2m_e}\nabla^2 + V_{ext}[\rho(\vec{r}_i)] + \int \frac{e^2\rho(\vec{r}_j)d^3\vec{r}_j}{|\vec{r}_i - \vec{r}_j|} + V_{XC}(\vec{r}_i)\right]\psi_i(\vec{r}_i) = \varepsilon_i\psi_i(\vec{r}_i) \qquad (2.28)$$

where the MOs $\psi_i(\vec{r}_i)$ are called *Kohn-Sham (KS) orbitals* and V_{XC} is the *exchange-correlation functional* which is the functional derivative of the exchange-correlation energy with respect to $\rho(\vec{r})$

$$V_{XC}(\vec{r}_i) = \frac{\delta E_{XC}[\rho(\vec{r})]}{\delta\rho(\vec{r})}. \qquad (2.29)$$

In the calculations the one-electron density is defined as

$$\rho(\vec{r}_i) = \sum_k n_k |\psi_k(\vec{r}_i)|^2, \qquad (2.30)$$

where n_k is the occupation number of kth molecular orbital, $\psi_k(\vec{r})$.

The knowledge of the ground-state density function $\rho_0(\vec{r})$ and the exchange-correlation energy functional $E_{XC}[\rho_0(\vec{r})]$ gives the exact energy. However, neither the exchange-correlation functional nor the exchange-correlation energy are known up to now. Therefore, a number of different approximations and strategies to the Kohn-Sham theory have been proposed to solve the equations.

The simplest approximation for solving the Kohn-Sham equations is the *Local Density Approximation* (LDA). The idea of the LDA consists in identifying the density locally as that of the *uniform electron gas*. Firstly, it is possible to present the exchange correlations energy in the following simple form

$$E_{xc}^{LDA}[\rho(\vec{r})] = \int \rho(\vec{r})\varepsilon_{xc}^{LDA}(\rho(\vec{r}))d\vec{r}. \qquad (2.31)$$

23

Here $\varepsilon_{xc}(\rho(\vec{r}))$ is the exchange-correlation energy per electron as a function of the density in a homogeneous electron gas which is known accurately for all densities from various approaches. The corresponding exchange-correlation functional depends only on $\rho(\vec{r})$ as:

$$V_{XC}(\vec{r}_i) = \rho(\vec{r})\frac{d\varepsilon_{xc}(\rho(\vec{r}))}{d\rho(\vec{r})} + \varepsilon_{xc}(\rho(\vec{r})) \,. \qquad (2.32)$$

Secondly, the exchange energy of an electron in a uniform electron gas may be given by Dirac's formula as

$$\varepsilon_x^{LDA}[\rho(\vec{r})] = -\frac{3e^2}{4}\left(\frac{3}{\pi}\right)^{1/3}\rho^{1/3}(\vec{r}) \qquad (2.33)$$

and

$$E_x^{LDA}[\rho(\vec{r})] = -\frac{3e^2}{4}\left(\frac{3}{\pi}\right)^{1/3}\int \rho^{4/3}(\vec{r})d\vec{r}. \qquad (2.34)$$

It should further be noted that, in contrast to the exchange energy ε_x, there is no simple functional form for the correlation energy ε_c, though accurate values of the correlation energy are known from various approaches such as quantum Monte-Carlo calculations (Ceperley and Alder, 1980, Vosko et al. 1980).

The majority of the molecules have an even number of electrons which are paired giving an overall singlet. These *closed-shell* systems can be described by the standard LDA procedure. In this case, each spatial KS orbital is occupied by two electrons with opposite spins α and β, i.e. the spatial orbital consists of two spin orbitals. The spin orbitals of the same spatial KS orbital have the same orbital energy. However, in particular magnetic systems with one or more unpaired electrons are *open-shell* systems. In this case, it is better to consider two systems of the α and β electrons separately with different spatial KS wavefunctions ψ_α and ψ_β, and the individual orbitals being considered as singly occupied. As a result, the α and β orbitals have different orbital energies and spatial parts. From these separate considerations of the spatial KS wavefunctions follow different electron density functions ρ_α and ρ_β. This "unrestricted" LDA method is called the *Local Spin Density Approximation* (LSDA). In this case, the total probability density is the sum of the individual spin densities $\rho = \rho_\alpha + \rho_\beta$ and the exchange energies are

$$E_x^{LSDA}[\rho_\alpha(\vec{r}),\rho_\beta(\vec{r})] = -\frac{3e^2}{4}\left(\frac{3}{2\pi}\right)^{1/3}\int\left(\rho_\alpha^{4/3}(\vec{r}) + \rho_\beta^{4/3}(\vec{r})\right)d\vec{r}, \qquad (2.35)$$

$$\varepsilon_x^{LSDA}[\rho_\alpha(\vec{r}),\rho_\beta(\vec{r})] = -\frac{3e^2}{4}\left(\frac{3}{2\pi}\right)^{1/3}\left(\rho_\alpha^{1/3}(\vec{r}) + \rho_\beta^{1/3}(\vec{r})\right). \qquad (2.36)$$

Generally, the L(S)DA yields good structural and ground-state properties for both molecules and solids. In practice, DFT calculations require even less efforts than HF calculations. Furthermore, increasing the size of the basis set gives a better description of the Kohn-Sham orbitals. The basis set requirements for DFT methods are similar to those for HF methods, since the DFT energy depends directly on the electron density. Another question is: are DFT methods *ab initio* or semi-empirical? *Ab initio* means that the method is based on

24

theory and any fitting parameters are absent. In this sense, LSDA methods are *ab initio*, because LSDA exchange energy does not contain any parameters, and the correlation functional is known accurately either as a tabulated function of the density or as analytical representations of this function.

Further approximation within LSDA is the non-local exchange-only or Xα approximation

$$U_{HF}^x(\vec{r}) = -\frac{3\alpha e}{2}\left(\frac{3\rho(\vec{r})}{\pi}\right)^{1/3}. \tag{2.37}$$

The corresponding Schrödinger equation is usually denoted as the *Hartree-Fock-Slater equation*

$$\left[-\frac{\hbar^2}{2m_e}\nabla^2 - \sum_I \frac{Z_I e^2}{|\vec{R_I}-\vec{r_i}|} + \int\frac{e^2\rho(\vec{r_j})d^3\vec{r_j}}{|\vec{r_i}-\vec{r_j}|} - \frac{3\alpha e^2}{2}\left(\frac{3\rho(\vec{r_i})}{\pi}\right)^{1/3}\right]\psi_i(\vec{r_i}) = \varepsilon_i\psi_i(\vec{r_i}). \tag{2.38}$$

Here, the first term is the kinetic energy of the electron, the second term is the attractive potential of the nuclei, the third term is the repulsive potential of the electron cloud of the whole system and the fourth term is the exchange potential. α is an exchange constant, which is chosen in our calculations $\alpha=0.7$. The simple expression for the exchange energy, Xα, in the HFS model drastically reduces computing time as compared with the ab initio HF model. Originally, the Xα method was developed for solids but has later been applied for molecules as well.

Another important question is: how one can judge whether the results of calculations are correct? The results of a theoretical model should always be compared with experiment. One direct link that can be made between the results of calculations and experimental measurements is a comparison of the calculated orbital energies with the measured ionization potential of a molecule (Anslyn 2006). It was shown (Luo et al, 2006) that in DFT, Koopman´s theorem for a large molecular system can be stated as follows: *the ionization energy of the system equals the negative of the highest occupied molecular orbital (HOMO) energy plus the Coulomb electrostatic energy of removing an electron from the system*, or equivalently, *the ionization energy of an N-electron system is the negative of the arithmetic average of the HOMO energy of this system and the lowest unoccupied molecular orbital (LUMO) energy of the (N-1)-electron system*.

2.3 Self-Consistent-Charge-Xα Method

The most efficient DFT method in the local spin-density approximation for calculating spectroscopic data is the spin-polarized self-consistent-charge (SCC)-Xα method (Grodzicki 1980, 1985).

The central point is the solution of the Kohn-Sham equations in order to find the one-electron particle density for the system under consideration. In the calculations it is defined as

$$\rho(\vec{r}) = \sum_k n_k |\psi_k(\vec{r})|^2, \qquad (2.39)$$

where n_k is the occupation number of kth molecular orbital, $\psi_k(\vec{r})$.

One may construct wavefunctions, $\psi_k(\vec{r})$, in any manner one deems reasonable, and one can judge the quality of the constructed wavefunctions by evaluation of the energy eigenvalues associated with each $\psi_k(\vec{r})$. The wavefunction with the lowest energy will be the most reliable and most likely be the best one to calculate other properties of the system. The choice of mathematical functions in order to construct a trial wavefunction is a typical question in mathematics. The suitable functions are called a *basis set* of the wavefunction.

Let us consider the choice of the trial wavefunction from the physical point of view. If an electron is in the vicinity of one nucleus of a molecule but far away from other nuclei then the interaction with other nuclei is negligible. Hence, near each nucleus, molecular electronic wavefunction must resemble an atomic orbital centered on that nucleus (Hinchliffe 2003). Such an assumption supposes an investigation of the approximate molecular wavefunction, ψ_k, as a linear combination of atomic orbitals, φ_i^I:

$$\psi_k(\vec{r}) = \sum_{I,i} c_{ik}^I \varphi_i^I(\vec{r} - \vec{R}_I), \qquad (2.40)$$

where the set of AO functions, $\{\varphi_i^I\}$, is the basis set. In equation (2.40), φ_i^I is i^{th} atomic orbital of Ith atom and c_{ik}^I is *expansion coefficient* which describes the contribution of atomic orbital φ_i^I in molecular orbital ψ_k. This approach is known as the *linear combination of atomic orbitals* (LCAO) approximation.

Moreover, as noted earlier, the choice of the most appropriate basis set is very important because it defines the accuracy and costs of the calculations. In general, three types of basis sets are used (Anslyn 2006):

(1) *All-electron*: all orbitals that are (partially) occupied in the atomic ground state, are included (e.g., *1s, 2s,* and *2p* for O);

(2) *Valence*: only valence orbitals are included (e.g., *2s* and *2p* for O or *3s* and *3p* for Si);

(3) *Extended*: extra functions beyond the all-electron or valence basis are added (e.g., *d* on C).

Further, it is possible to write equation (2.38) in the general form as:

$$\hat{H}\psi_k(\vec{r}) = \varepsilon_k \psi_k(\vec{r}). \qquad (2.41)$$

Using here equation (2.40) we have

$$\sum_{I,i} c_{ik}^I \hat{H}\varphi_i^I(\vec{r} - \vec{R}_I) = \varepsilon_k \sum_{I,i} c_{ik}^I \varphi_i^I(\vec{r} - \vec{R}_I). \qquad (2.42)$$

Multiplying this equation by $\varphi_j^J(\vec{r} - \vec{R}_J)$ and integrating over $\vec{r}$ leads to the *secular equation*

$$\sum_{I,i}\left(H_{ij}^{IJ} - \varepsilon_k S_{ij}^{IJ}\right) c_{ik}^I = 0, \qquad (2.43)$$

26

with the matrix elements

$$H_{ij}^{IJ} = \int \varphi_j^J(\vec{r} - \vec{R}_J) \left[-\frac{\hbar^2}{2m_e} \nabla^2 + v_{mol}(\vec{r}) \right] \varphi_i^I(\vec{r} - \vec{R}_I) d^3r \qquad (2.44)$$

and

$$S_{ij}^{IJ} = \int \varphi_j^J(\vec{r} - \vec{R}_J) \varphi_i^I(\vec{r} - \vec{R}_I) d^3r. \qquad (2.45)$$

In equation (2.44), the matrix elements H_{ij}^{IJ} are the Hamitonian matrix elements and $v_{mol}(\vec{r})$ consists of the classical Coulomb potential and the exchange potential

$$v_{mol}(\vec{r}) = v_{cb}[\rho(\vec{r})] + v_x[\rho(\vec{r})]. \qquad (2.46)$$

In equation (2.45), S_{ij}^{IJ} are overlap integrals, describing the degree of orthogonality, or more generally, the overlap of two orbitals of interacting atoms. For normalized atomic orbitals the values of overlap integrals lie between −1 and +1. If orbitals are orthogonal the overlap integral is zero, and otherwise the overlap integral exponentially approaches zero if the two nuclei are very far apart. A consideration of the overlap integral is very important during the determination of the strength of bonding interactions; in particular, strong bonds result from significant overlap.

The efficiency of a DFT method is defined through the art of calculation of the integrals (2.44). Here, a three-dimensional numerical integration would be necessary without subsequent simplifications. In SCC-Xα approximation, the calculations are reduced with a help of model simplifications and transformations to a maximal one-dimensional numerical integration.

In the first step we define an atomic Hamiltonian (Grodzicki 1980)

$$H_{at}^I = -\frac{\hbar^2}{2m_e} \Delta + v_{at}^I(\vec{r} - \vec{R}_I), \qquad (2.47)$$

which describes an atom I at position $\vec{R}_I$. Substitution of equations (2.46) and (2.47) in the matrix elements (2.44) of the Hamiltonian gives

$$H_{ij}^{IJ} = \int \varphi_j^J(\vec{r} - \vec{R}_J) \left[\frac{1}{2}(H_{at}^I + H_{at}^J) + v_{nb}^{IJ}(\vec{r}) \right] \varphi_i^I(\vec{r} - \vec{R}_I) d^3r =$$

$$= \frac{1}{2}(\varepsilon_i^I + \varepsilon_j^J) S_{ij}^{IJ} + \int \varphi_j^J(\vec{r} - \vec{R}_J) v_{nb}^{IJ}(\vec{r} - \vec{R}_I) \varphi_i^I(\vec{r} - \vec{R}_I) d^3r, \qquad (2.48)$$

where the *neighbor potential*

$$v_{nb}^{IJ}(\vec{r}) = v_{mol}(\vec{r}) - \frac{1}{2}\left(v_{at}^I(\vec{r} - \vec{R}_I) + v_{at}^J(\vec{r} - \vec{R}_J) \right) \qquad (2.49)$$

and the *orbital energies* of the Ith atom or ion

$$\varepsilon_i^I = \varepsilon_{i0}^I + \varepsilon_{i1}^I Q_I + \varepsilon_{i2}^I Q_I^2 \qquad (2.50)$$

are assumed as a quadratic function of the *effective charge*

$$Q_I = Z_I^c - N_I^{val}. \tag{2.51}$$

In equations (2.50) and (2.51), $\varepsilon_{i0}^I, \varepsilon_{i1}^I$ and ε_{i2}^I are numerical parameters which are derived from atomic calculations, Z_I^c and N_I^{val} are the core charge (e.g. +4 for Si, +6 for O) and the number of valence electrons assigned to atom I in the molecule, respectively (Mulliken 1955, Grodzicki 1980).

It is important to note that the substitution of (2.47) in (2.44) and evaluation of the Hamilton matrix elements (2.48) using orbital energies (2.50) avoids the direct calculation of the kinetic energy which is proportional to the second derivative of the wavefunction. In turn, second derivative has large contribution from core regions where the radial part has oscillations and, therefore, required extended basis set for calculation of the kinetic energy. Because, the kinetic energy implicitly calculated, we can use valence basis set and simplify calculations.

In the second step the approximated molecular electron density within the frame of the SCC-Xα method is

$$\rho(\vec{r}) = \sum_K \rho_{at}^K(|\vec{r} - \vec{R}_K|). \tag{2.52}$$

Because, the Coulomb potential of both the nuclei and electrons is a linear function of the particle density the molecular potential can be written as a sum of atomic Coulomb potentials

$$v_{mol}(\vec{r}) = \sum_K v_{cb}^K(|\vec{r} - \vec{R}_K|) + v_{xc}[\rho(\vec{r})]. \tag{2.53}$$

where atomic Coulomb potentials are

$$v_{cb}^K(\vec{r}) = -\frac{Z_K e^2}{|\vec{r} - \vec{R}_K|} + \int \frac{e^2 \rho_{at}^K(\vec{r}\prime) d^3 r\prime}{|\vec{r} - \vec{R}_K - \vec{r}\prime|}. \tag{2.54}$$

Shifting the origin of the matrix-elements to the Ith atom yields

$$H_{ij}^{IJ} = \int \varphi_j^J(\vec{r} - \vec{R}_{JI})\left[H_{at}^I + v_{nb}^I(\vec{r} - \vec{R}_I)\right]\varphi_i^I(\vec{r}) d^3 r =$$

$$= \varepsilon_i^I S_{ij}^{IJ} + \int \varphi_j^J(\vec{r} - \vec{R}_{JI})v_{nb}^I(\vec{r} - \vec{R}_I)\varphi_i^I(\vec{r}) d^3 r. \tag{2.55}$$

with

$$\vec{R}_{JI} = \vec{R}_J - \vec{R}_I, \tag{2.56}$$

Thus, substituting (2.52)-(2.55) in the Hamiltonian matrix element (2.56), it can be written as

$$H_{ij}^{IJ} = \int \varphi_j^J(\vec{r} - \vec{R}_{JI})\{H_{at}^I + \sum_{K \neq I} v_{cb}^K(|\vec{r} - \vec{R}_{KI}|) + v_x[\rho(\vec{r})] - v_x[\rho_{at}^I(\vec{r})]\}\varphi_i^I(\vec{r}) d^3 r. \tag{2.57}$$

In equation (2.57) different types of integrals occur as listed in Table 2.1.

Since the one-center integrals are larger by approximately an order of magnitude than the two- and three-center integrals, the one-center integrals require highest accuracy in the calculations whereas requirements for the evaluation of multi-center integrals are considerably less restrictive. Hence, the basic idea within the framework of the SCC-Xα method is the treatment of the one-center and multi-center integrals on different levels of accuracy.

Large basis sets, accurate pseudopotentials, nonlocal exchange functionals are mainly needed for one-center integrals, and consequently they are derived from numerical atomic all-electron calculations with a semi-local (gradient corrected) exchange-correlation functional. In turn, the multi-center integrals are evaluated with minimal basis sets, simple core potentials and the local exchange-only functional, Xα (Grodzicki 1985).

Table 2.1 Types of integrals in Hamiltonian matrix elements

$I=J$	$\int \varphi_j^I(\vec{r}) H_{at}^{II} \varphi_i^I(\vec{r}) d^3r$	1-center atomic
$K \neq I=J$	$\int \varphi_j^I(\vec{r}) v_{cb}^K(\lvert\vec{r} - \vec{R}_{KI}\rvert) \varphi_i^I(\vec{r}) d^3r$	Coulomb crystal-field
	$\int \varphi_j^I(\vec{r}) v_x[\rho(\vec{r})] \varphi_i^I(\vec{r}) d^3r$	exchange crystal-field
$I \neq J$	$\int \varphi_j^J(\vec{r} - \vec{R}_{JI}) \varphi_i^I(\vec{r}) d^3r$	overlap
$I \neq J$	$\int \varphi_j^J(\vec{r} - \vec{R}_{JI}) H_{at}^{II} \varphi_i^I(\vec{r}) d^3r$	resonance
$K=I \neq J$	$\int \varphi_j^J(\vec{r} - \vec{R}_{JI}) v_{cb}^I(\vec{r}) \varphi_i^I(\vec{r}) d^3r$	2-center Coulomb
$K \neq I \neq J$	$\int \varphi_j^J(\vec{r} - \vec{R}_{JI}) v_{cb}^K(\lvert\vec{r} - \vec{R}_{KI}\rvert) \varphi_i^I(\vec{r}) d^3r$	3-center Coulomb
$I \neq J$	$\int \varphi_j^J(\vec{r} - \vec{R}_{JI}) v_x[\rho(\vec{r})] \varphi_i^I(\vec{r}) d^3r$	2- and 3-center exchange

2.4 Evaluation of Multi-Center Integrals

First, it is possible to separate atomic wavefunctions into radial and angular parts, because of the spherical symmetry of atoms (Grodzicki 1985)

$$\varphi_{nlm}(\vec{r}) = R_{nl}(r)Y_{lm}(\theta, \phi), \qquad (2.58)$$

where n, l and m are quantum numbers (principal, orbital and magnetic). The angular part,

$$Y_{l,m}(\theta, \phi) = (-1)^m \left(\frac{2l+1}{4\pi} \cdot \frac{(l-m)!}{(l+m)!}\right)^{1/2} P_l^m(\cos\theta)e^{im\phi}, \qquad (2.59a)$$

is the same for defined l and m in any atom and associated with Legendre polynomials,

$$P_l^m(\cos\theta) = (-l)^m \left(\frac{2}{\pi l \sin\theta}\right)^{1/2} \cos\left[\left(l + \frac{1}{2}\right)\theta - \frac{\pi}{2} + \frac{m\pi}{2}\right] + O(l^{-1/2}), \quad (2.59b)$$

(Edmonds, 1964), whereas only the radial part, $R_{nl}(r)$, is specific for a particular atom.

Therefore, it is necessary to choose the mathematical form for the radial part of atomic orbitals in the basis set. It is logical to propose the radial parts of hydrogenic wavefunctions for this purpose, based on the fact that the Schrödinger equation for the hydrogen atom is solved exactly. The problem is that the calculation of the corresponding integrals is very complicated. Therefore, the radial part of is chosen as the *Slater-type orbital*

$$R_{nl}(r) = a_{nl}r^{n-1}e^{-\zeta_{nl}r}. \qquad (2.60)$$

In equation (2.60), a_{nl} is the normalization constant and ζ_{nl} is the orbital- and charge-dependent parameter

$$\zeta_{nl} = \zeta_{nl}(Q) = \zeta_{nl\,0} + \zeta_{nl\,1}Q. \qquad (2.61)$$

In practice, equation (2.60) means that the stronger electron is bound in the atom (the larger $|\varepsilon_{nl}|$), the more localized is the corresponding wavefunction (the larger ζ_{nl}) for s- and p-orbitals, whereas an exception exists for $3d$ and $4f$ orbitals.

Since the direct calculation of the kinetic energy integrals is avoided, it is possible to utilize a minimal basis set for evaluating the multi-center integrals. Using equations (2.58) and (2.60), we can represent every valence orbital with the single *Slater-type orbital (STO)*

$$\varphi_{nlm}(\vec{r}) = a_{nl}r^{n-1}e^{-\zeta_{nl}r}Y_{lm}(\theta, \phi). \qquad (2.62)$$

Again, it should be emphasized that the STOs describe atomic orbitals correctly only in the bonding region and not in the core region where the radial part has oscillations. It should further be noted that closed shells are energetically very stable and atomic electron density is

$$\rho(\vec{r}) = \sum_{nl} x_{nl} \sum_m |\varphi_{nlm}(\vec{r})|^2 = \sum_{nl} x_{nl} R_{nl}^2(r), \qquad (2.63)$$

where x_{nl} is the occupation number of all atomic orbitals $\varphi_{nlm}(\vec{r})$ with equal nl. As a result, atomic electron densities can be approximated by (Grodzicki 1980)

$$\rho(\vec{r}) = \sum_i X_i e^{-\eta_i r}, \tag{2.64}$$

where η_i is a numerical charge-dependent parameter which describes the decay of the valence density

$$\eta_i = \eta_i(Q) = \eta_{i0} + \eta_{i1} Q. \tag{2.65}$$

Second approximation of the SCC-Xα method is the core-valence separation of the atomic electron density:

$$\rho(\vec{r}) = \rho_c(\vec{r}) + \rho_{val}(\vec{r}) = \sum_K \rho_c^K(|\vec{r} - \vec{R}_K|) + \sum_K \rho_{val}^K(|\vec{r} - \vec{R}_K|). \tag{2.66}$$

Taking this approximation into account and assuming that the valence electron density is negligible in the core region and the core electron density is equal zero in the valence region, the exchange potential can be also separated:

$$v_x[\rho(\vec{r})] = v_x[\rho_c(\vec{r})] + v_x[\rho_{val}(\vec{r})]. \tag{2.67}$$

Within the Xα approximation, the exchange potential can thus be written as:

$$v_x[\rho(\vec{r})] = -3\alpha \left(\frac{3}{\pi}\right)^{1/3} \left\{ \left(\rho_c(\vec{r})\right)^{1/3} + \left(\rho_{val}(\vec{r})\right)^{1/3} \right\}. \tag{2.68}$$

In the third step, it is assumed that there is no overlap between the core densities at different sites, so that

$$v_{xc}[\rho_c(\vec{r})] = -3\alpha \left(\frac{3}{\pi}\right)^{1/3} \sum_K \left\{ \rho_c^K(|\vec{r} - \vec{R}_K|) \right\}^{1/3}. \tag{2.69}$$

In the fourth assumption, both densities are represented by a single exponential

$$\rho^K(\vec{r} - \vec{R}_K) = \frac{N_K \eta_K^3}{8\pi} e^{-\eta_K |\vec{r} - \vec{R}_K|}, \tag{2.70}$$

where $\eta_{val,K} = \eta_{val,K}(Q)$ is the parameter (2.65) describing the decay of the valence density obtained from atomic calculations. This assumption has several substantial advantages, such as simple representation of the Coulomb potential

$$\frac{v_{cb}^K(|\vec{r} - \vec{R}_K|)}{e^2} = -\frac{Q_K}{|\vec{r} - \vec{R}_K|} - N_K \left(\frac{1}{|\vec{r} - \vec{R}_K|} + \frac{\eta_K}{2}\right) e^{-\eta_K |\vec{r} - \vec{R}_K|}, \tag{2.71}$$

and separation of the Coulomb potential in radial and angular parts using the expansion formula (Barnett and Coulson, 1951)

$$|\vec{r} - \vec{R}|^n e^{-\eta |\vec{r} - \vec{R}|} = 4\pi \sum_{lm} (-1)^{n+l+1} \beta_l^{(n)}(\eta; r_<, r_>) Y_{lm}^*(\hat{R}) Y_{lm}(\hat{r}), \tag{2.72}$$

where $r_< = min(r, R)$, $r_> = max(r, R)$, $\beta_l^{(n)}$ is the radial coefficient, calculated from simple recurrence relations (Barnett and Coulson, 1951) and Y_{lm} is the spherical harmonic.

Thus, the valence Coulomb potential can be written as

$$v_{cb}^K\left(|\vec{r} - \vec{R}_K|\right) = 4\pi \sum_{lm} v_{cb,l}^K(r, R_K) Y_{lm}^*\left(\hat{R}_K\right) Y_{lm}(\hat{r}), \qquad (2.73)$$

where

$$v_{cb,l}^K(r, R_K) = -\frac{2Q_K}{2l+1} \cdot \frac{r_<^l}{r_>^{l+1}} - (-1)^l N_K (2\beta_l^{(-1)} - \eta_K \beta_l^{(0)}). \qquad (2.74)$$

The evaluation of the molecular exchange potential is somewhat more complicated. With respect to a site $\vec{R}_I$ the exchange potential is

$$v_x^I[\rho(\vec{r})] = -\alpha'\left(\sum_K N_K \eta_K^3 e^{-\eta_K|\vec{r}-\vec{R}_{KI}|}\right)^{1/3}, \qquad (2.75)$$

where $\alpha' = \frac{3}{2}\left(\frac{3}{\pi^2}\right)^{1/3} \alpha = 1.00855\alpha = 0.706$. Equation (2.75) can be expressed through the one-center expansion formula (2.72) as

$$v_x^I[\rho(\vec{r})] = -\alpha'\left\{P_I(r) - 4\pi \sum_{K \neq I} \sum_{l(>0)m} (-1)^l N_K \eta_K^3 \, \beta_l^{(0)}(\eta_K; r_<, r_>) Y_{lm}^*\left(\hat{R}_{KI}\right) Y_{lm}(\hat{r})\right\}^{1/3},$$

$$(2.76)$$

where

$$P_I(r) = N_I \eta_I^3 e^{-\eta_I r} - \sum_{K \neq I} N_K \eta_K^3 \, \beta_0^{(0)}(\eta_K; r_<, r_>). \qquad (2.77)$$

Equation (2.76) can be approximated using a Taylor expansion with termination after the first (linear) term as:

$$v_x^I[\rho(\vec{r})] = -\alpha'\left\{P_I^{1/3}(r) - 4\pi \sum_{K \neq I} \sum_{l(>0)m} \tilde{\rho}_l^K(r, R_{KI}) Y_{lm}^*\left(\hat{R}_{KI}\right) Y_{lm}(\hat{r})\right\} \qquad (2.78)$$

where

$$\tilde{\rho}_l^K(r, R_{KI}) = N_K \eta_K^3 \frac{(-1)^l \beta_l^{(0)}(\eta_K; r_<, r_>)}{3P_I^{2/3}(r)}. \qquad (2.79)$$

Hence, the exchange potential from the neighbours (2.78) and the Coulomb potential (2.73) exhibit the same angular dependence simplifying the calculations considerably. Consequently, it is possible to represent the molecular potential from the neighbours as one-centre series expansion with respect to the site $\vec{R}_I$ as (Grodzicki 1980):

$$v_{nb}^I(\vec{r}) = v_0^I(r) + 4\pi \sum_{K \neq I} \sum_{l(>0)m} v_l^K(r, R_{KI}) Y_{lm}^*\left(\hat{R}_{KI}\right) Y_{lm}(\hat{r}), \quad (2.80)$$

where

$$v_0^I(r) = \sum_{K \neq I} v_0^K(r, R_{KI}) - \alpha'\left\{P_I^{1/3}(r) - [\rho_{at}^I(r)]^{1/3}\right\}, \qquad (2.81)$$

and

32

$$v_l^K(r, R_{KI}) = v_{cb,l}^K(r, R_{KI}) - \alpha' \tilde{\rho}_l^K(r, R_{KI}). \qquad (2.82)$$

It is usually sufficient to terminate the mathematically infinite l-summation with $l \approx 6$.

It should further be noted that the matrix elements in (2.44) are functions of the effective charges of the atoms in the molecule, equations (2.50), (2.51), (2.61) and (2.65). The values of these atomic parameters which are used in our calculations are given in Appendix 1.

The number of valence electrons, N_l^{val}, is determined from the overlap population

$$N_{ij}^{IJ} = P_{ij}^{IJ} \cdot S_{ij}^{IJ}, \qquad (2.83)$$

where S_{ij}^{IJ} is the overlap integral (2.45) between two AOs and P_{ij}^{IJ} is the bond-order-matrix

$$P_{ij}^{IJ} = \sum_k x_K \, c_{ik}^I c_{jk}^J, \qquad (2.84)$$

which describes the distribution of the electrons over the AOs of the basis set. The product (2.83) can be described as the number of electrons in the atomic orbital i at atom I and j at atom J, which contribute to the bond between I and J, if $I \neq J$. Hence, if $I = J$ and $i = j$, the product (2.83) describes the net population (Grodzicki et al. 1983) of the ith orbital at atom I

$$N_i^I = N_{ii}^{II} + \sum_{I \neq J, j} \frac{a_i^I}{a_j^J + a_i^I} \cdot N_{ij}^{IJ}, \qquad (2.85)$$

where a_i^I are weighting factors which are determined by numerical integration (Grodzicki 1985).

The number of valence electrons at atom I is then

$$N_I^{val} = \sum_i N_i^I. \qquad (2.86)$$

The difference of orbital occupation numbers between two following iterations (n) and $(n+1)$ serves as SCC-Xα convergence criterion

$$\max_{I,i} \left| N_i^{I,(n+1)} - N_i^{I,(n)} \right| < \varepsilon. \qquad (2.87)$$

The structure of the self-consistent procedure for solving the HFS equation in the SCC-Xα method consists in:

1.   initial guess of "reasonable" orbital occupation numbers $N_i^{I,(0)}$ for all (n,l) AOs of the basis set;

2.   calculation of the Hamiltonian matrix elements $H_{ij}^{IJ(0)}$, overlap integrals $S_{ij}^{IJ(0)}$ and solving secular equations

$$\left| H_{ij}^{IJ(0)} - \varepsilon_k^{(0)} S_{ij}^{IJ(0)} \right| = 0; \qquad (2.88)$$

3.   next, eigenvalues $\varepsilon_k^{(0)}$ and eigenvectors $c_{ik}^{I(0)}$ are obtained from the equation (2.88);

4. using the eigenvectors $c_{ik}^{I,(0)}$ a new set of orbital occupation numbers $N_i^{I,(1)}$ are determined as starting values for the next iteration;

5. repetition of these steps, until the procedure has converged, i.e. the input and output occupation numbers differ less than a given threshold,

$$\max_{I,i} \left| N_i^{I,(n+1)} - N_i^{I,(n)} \right| < \varepsilon. \qquad (2.89)$$

Undoubtedly, the SCC-Xα method has substantial advantages and, of course, disadvantages in comparison with other DFT-methods. Factorization of the molecular potential into a radial part and spherical harmonics enables the analytical integration of the angular parts of the Hamiltonian matrix elements. Thus, all symmetry properties of the system are automatically preserved. Next, one-dimensional numerical integrations are required only for the three-centre Coulomb, and two- and three-centre exchange integrals leading to a substantial reduction of CPU time (central processing unit time) and memory space. In addition, CPU time is minimized due to using a minimal basis set for evaluating all multi-centre integrals.

Disadvantages of the method are simplifying model assumptions for the atomic electron densities, separate determination of atomic and molecular data (no "black-box" procedure) and restricted flexibility in the systematic improvements. Nevertheless, minimization of CPU time and memory space, and reliable results of the calculations makes this method useful and applicable for real mineralogical crystal structures and other large and complex systems.

2.5 Cluster Construction

The common approach for electronic structure calculations on extended periodic systems relies on band structure calculations. Since, however, both hyperfine parameters and magnetic interactions in insulators are genuinely local in nature, the use of cluster models offers a viable alternative. Such a cluster approach has a number of obvious advantages among which the most important are the independence of the number of atoms per unit cell and the possibility of modeling solid solutions and local cationic disorder without additional efforts. Moreover, complex arithmetics can be avoided in cluster calculations, which is essential when dealing with larger systems. The most severe objection against cluster calculations may be that they cannot account for effects arising from the extended nature of the solid. Indeed, if the hyperfine parameters and magnetic interactions were dominated by collective effects, cluster models will fail to provide reliable results. Both experimental and theoretical experience, however, have supplied convincing evidence for the absence of collective effects in hyperfine parameters. Next, in magnetic insulators, the magnetic centres interact through diamagnetic anionic ligands, as is the case in the iron-bearing silicates. The interacting spins can usually be assumed in very good approximation to be localized at centres A and B. Under this assumption, the Heisenberg Hamiltonian (Section 3.3) provides an appropriate basis for describing the magnetic properties. It is important to emphasize that the calculation of the energies for one system have to be carried out with respect to the same geometry.

It has been shown (Grodzicki and Lebernegg 2010) that small, highly charged model clusters like $(FeO_n)^{n-}$ including just the first coordination sphere of the Fe ion under consideration are not a suitable approximation for studying the electronic properties of Fe-bearing compounds. According to previous experience (Lougear et al. 2000, Grodzicki and Amthauer 2000, Grodzicki et. al. 2001, 2003, Weber et al. 2007, Zherebetskyy et al. 2010) with cluster calculations on various Fe-bearing silicates and phosphates the key issue of such calculations is size convergence, i.e., to warrant that the calculated spectroscopic data are converged with respect to the cluster size. In general, size converged result for optical transitions, hyperfine parameters and magnetic properties are obtained if all coordination polyhedra of the cations bonded to the oxygen atoms of the first coordination sphere of the central Fe ion are included. However, the large number of terminal anions of these added coordination polyhedra which are unsaturated in the sense that the next shell of cations is missing will again lead to an inappropriate description of the electronic properties. First of all, the oxidation number –2 of oxygen results in a large negative cluster charge that causes convergency problems ("embedding problem") if not compensated by a Madelung-type potential, e.g., of appropriately distributed point charges. Secondly, omitting the cations of the next shell bonded to the terminal oxygens produces unsaturated $O(2p)$-lone-pair orbitals that are in the same energy range as the $3d$-orbitals of Fe and generally cause convergency problems, as well. Actually, both the Madelung potential and the bonding to the outer cations will lead to a stabilization of the $O(2p)$-orbitals and, therefore, would remove these convergency problems. This stabilization can be attained by replacing some or all of the terminal oxygens with fluorine atoms or by omitting them, at all, if the metal-oxygen distances are sufficiently large. Such a procedure not just shifts the $2p$-orbitals into the energy range typical for saturated $O(2p)$-orbitals but also leads to a significant reduction of the cluster charge, and generally to neutral clusters circumventing, thus, the embedding problem. Alternatively, further coordination polyhedra can stepwise be added in such a way that the local symmetry of the metal site under consideration is preserved, though at the expense of CPU time and memory space. As former experience has shown, a combination of both strategies, i.e. adding further polyhedra and replacing incompletely coordinated oxygens with fluorine, has to be pursued and leads to a hierarchy of clusters of increasing size that may serve as a control for size convergence. Clusters that are size-converged with regard to calculated spectroscopic data usually contain between 100 and 150 atoms, and represent the correct surroundings of the central Fe ion within a sphere with radius of about 5 Å. Finally, transition metal ions in more distant polyhedra may be substituted by main group metal ions of the same oxidation state, e.g. Mg instead of Fe^{2+} and Al instead of Fe^{3+}, because the resulting changes in the computed spectroscopic data are within the error margins expected due to the general theoretical approximations. Cluster calculations based on this strategy are capable to supply a comprehensive description of the physical and chemical properties of Fe-bearing extended systems provided that size-converged clusters are being used.

Further, in order to derive and explain the spin structure of a mineral, clusters of increasing size around interacting magnetic centres have to be constructed based on the possible exchange pathways between the centres (Section 3.3). In these clusters, the interacting spins of the valence spin-up and spin-down orbitals in the magnetic centres have to be oriented in accordance with the ferromagnetic and antiferromagnetic configurations and respective total energies have to be calculated. From these energies, the respective Heisenberg coupling constant J can be calculated using equation (3.46).

However, the carrying out of these calculations for the determination of the Heisenberg coupling constant is very problematic, because the crystal structure data for minerals at temperatures below the Néel temperature frequently are not yet available. Therefore, we have to use the crystal structure data for lowest available temperatures from the American Mineralogist Crystal Structure Database (AMCSD). Additionally, the strong delocalization of the magnetic orbitals over the various magnetic centres prevents sometimes the convergency and arriving at the electronic ground state.

Chapter 3

PHYSICAL PROPERTIES

As described in the preceding chapter, the calculations in the frame of the DFT permit to solve problems of electronic structure of molecules and of condensed matter. In order to assess the reliability of both the theoretical approach and the suitability of model clusters, computed spectroscopic data are compared with experimental results. Therefore, in this chapter the corresponding experimental methods are briefly being discussed.

3.1 Optical Absorption

One of the common experimental methods for determining the electronic structure of materials is *optical absorption spectroscopy* covering the range of the electromagnetic spectra from the near ultraviolet (UV: 190-380 nm) through the visible region (Vis: 380-750 nm) to the near-infrared (NIR: 750-2500 nm). Accordingly, absorption spectra are closely correlated with colour, the distribution of chemical elements over different sites in a mineral, thermodynamic properties etc. The literature on mineral sciences emphasises the great importance of absorption spectroscopy as a powerful experimental method in studying Earth's minerals, synthetic compounds as well as extraterrestrial materials as moon-samples, meteorites or cosmic dust (Burns 1993).

Since the absorption of electromagnetic radiation of certain energies corresponds to electronic transitions between stationary electronic states, understanding the electronic structure and properties is essential for the interpretation of absorption spectra (Wildner et al. 2004). The first theoretical attempt to interpret absorption spectra of transition metal bearing solids was the *crystal field theory* (CFT) (Bethe 1929) describing the splitting pattern of the *d*-orbitals of a transition metal ion in a crystalline environment. The distribution of electrons over atomic orbitals, yielding a so-called electron configuration, follows two principles. The first one is the *Pauli exclusion principle*, which states that no more than one electron occupies a nondegenerate spin orbital. The second enables the determination of the ground state of atoms with open shells and is known as *Hund's rules*.

Since this thesis deals with Fe-bearing silicates, where the corresponding electronic transitions can be described using the CFT, the subsequent discussion will focus on Fe as a representative *transition metal element* with a partly filled *d*-shell. In the ground state the full iron atom has the electronic configuration of $3d^6 4s^2$. Generally, in chemical compounds some of the outer electrons are removed. The most common oxidation states for iron are the ferrous

(Fe^{2+}) and ferric (Fe^{3+}) iron with the electronic configurations $3d^64s^0$ and $3d^54s^0$, respectively. Most of the iron bearing minerals contain iron ions in the high-spin state. According to Hund's rules ferrous and ferric iron have $3d_\uparrow^5 3d_\downarrow^1$ (^{5}D) and $3d_\uparrow^5 3d_\downarrow^0$ (^{6}S) spin states,

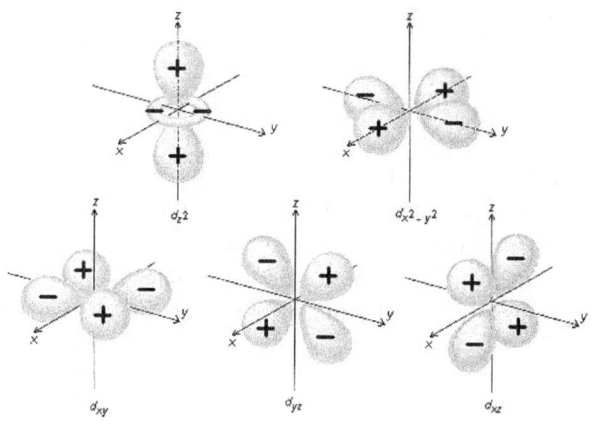

Figure 3.1 Angular distribution probabilities for electrons in five d-orbitals.

respectively. In the isolated ion the five $3d$-orbitals (Fig. 3.1) are energetically identical, or degenerate. When the ion is incorporated in a crystal, for example in an ideal octahedral or dodecahedral coordination, these five $3d$-orbitals of the transition metal are no longer degenerate but split into two groups, t_{2g} and e_g. According to CFT, the $3d$-orbitals are split due to the repulsion of the d electrons by the non-spherical electrostatic field of the negatively charged anions. The t_{2g} group (denoted by "t" for triply degenerate orbitals and by "g" for even with regard to inversion) consists of the d_{xy}, d_{yz} and d_{xz} orbitals, which have lobes oriented between the Cartesian axes. The e_g group consists of the two (denoted by "e" for doubly degenerated) $d_{x^2-y^2}$ and d_{z^2} orbitals, with their lobes directed along the Cartesian axes.

In case of an ideal octahedron with anions on the coordination axes, the lobes of the e_g orbitals are oriented directly towards the ligands and, therefore, electrons in these two orbitals are more strongly repelled than those in the three t_{2g} orbitals with lobes oriented between the ligands. In ideal rhombic dodecahedral coordination (cube) with anions on the space diagonals, there is the opposite case: electrons of the t_{2g} orbitals are repelled by the ligands to a greater extent than those in the two e_g orbitals. The energy separation between the t_{2g} and e_g orbitals is denoted as the *crystal field splitting*, Δ_i, where the subscript i refers to the polyhedral type, for example Δ_{oct} and Δ_{dod} for the octahedral and dodecahedral case, respectively. In octahedral symmetry, the crystal field splitting parameter derived from the point charge model is (Lever 1984, Burns 1993, Andrut et al. 2004):

$$\Delta_{oct} = \frac{5}{3} \frac{q\langle r^4 \rangle}{R^5} \qquad (3.1)$$

where q is the charge of the ligands, R is the metal-ligand distance and $\langle r^4 \rangle$ is the expectation value of the fourth power of r with respect to the $3d$ orbital.

Further, in a crystal field of distorted polyhedra with sufficiently low symmetry the degeneracy of the $3d$-orbitals is completely removed (Fig. 3.2). Generally, the splitting pattern depends on the symmetry and chemical type of the surrounding ligands.

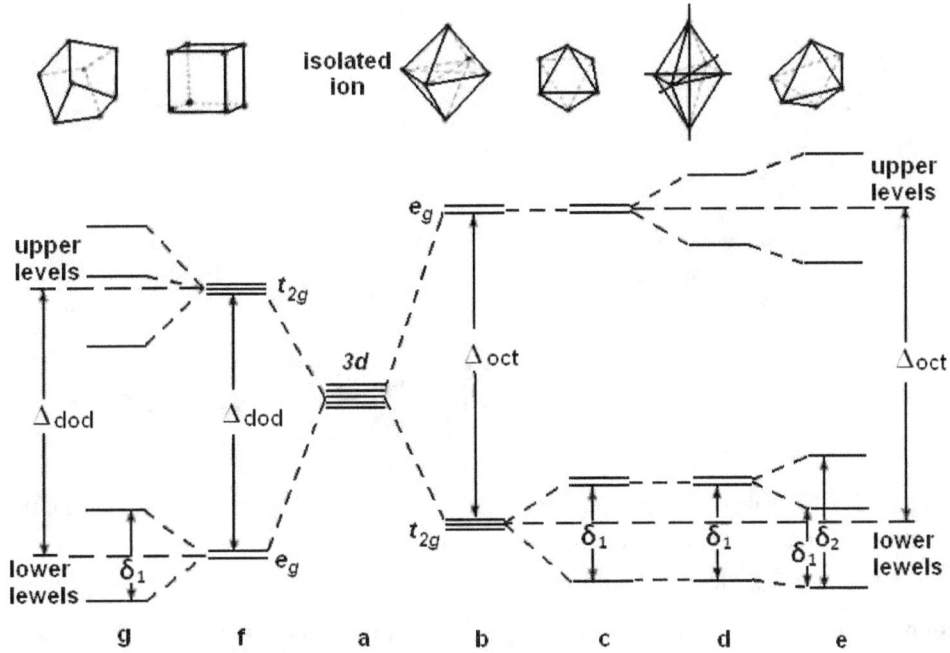

Figure 3.2 Splitting pattern of the $3d$ orbital energies for a transition metal ion in different environments: **(a)** - free ion, **(b)** - ideal octahedron, **(c)** - trigonally compressed octahedron, **(d)** - tetragonally elongated octahedron, **(e)** – completely unsymmetrical octahedron, **(f)** - ideal rhombic dodecahedron, **(g)** - distorted rhombic dodecahedron.

Different splittings of the $3d$ orbitals cause absorption bands at different energies in the UV-Vis-NIR region. This allows identifying the type and symmetry of the surrounding ligands and, in addition, the oxidation state and spin-state of the transition metal ion. Therefore, in order to understand features of the splitting pattern of the $3d$ orbital energies, we carried out calculations for uniform expansion of the high-spin Fe^{2+} cation in octahedral oxygen environment and various distortions in this system.

The splitting between the t_{2g} and e_g orbitals decreases with increasing Fe–O distances in the ideal octahedron (Fig. 3.3). An approximation of the t_{2g}–e_g splitting according to CFT by power function R^{-5} multiplied with an appropriate scaling factor exhibits considerable deviations (Fig. 3.3 a). Moreover, a best approximation by the fifth order inverse polynomial, $\Delta_{CF} = -5580R^{-5} + 14501R^{-4} - 14416R^{-3} + 6994R^{-2} - 1669R^{-1} + 157.4$ contains considerable contribution of additional terms, and a best-fit by the power function $\sim R^{-n}$

yields n=7.71 (Fig. 3.3 b). Both approximations are not predicted in the frame of CFT giving $\Delta_{CF} \sim R^{-5}$. Furthermore, the subsequent analysis convincingly demonstrates that there is no general theoretical justification for a strict dependence of the inverse fifth degree.

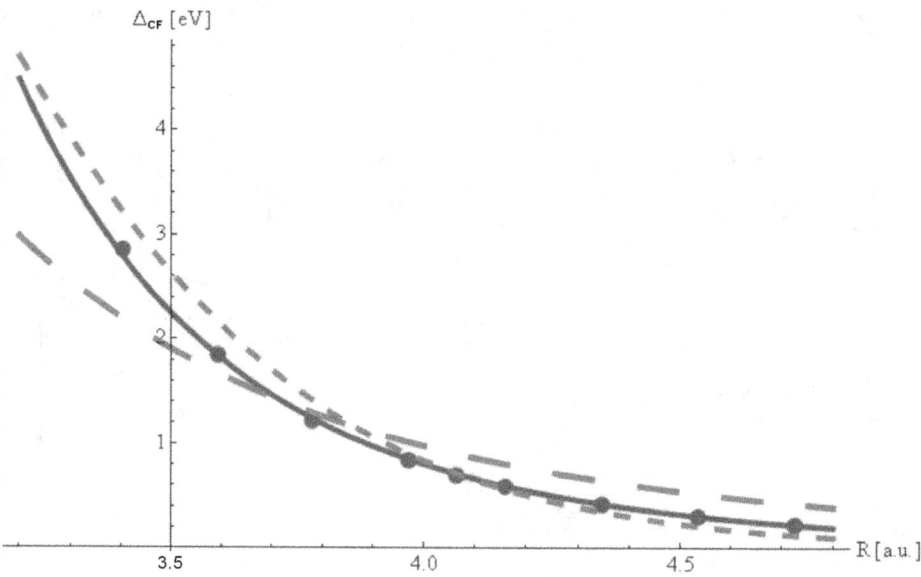

Figure 3.3 The calculated splitting (dots) between the t_{2g} and e_g orbitals for octahedral symmetry with approximated trend-lines:

a) CFT, $\Delta_{CF} = \frac{1000}{R^5}$, long-dashed (red) line;

b) best-fit by the power function, $\Delta_{CF} = \frac{35076}{R^{7.71}}$, solid (dark-blue) line;

c) theoretical approximation from equation (3.4) with ζ=2.16, ΔH_{sd}= –20eV and ΔH_{pd}= –3eV, short-dashed (light-blue) line (Lebernegg 2010).

This incapability and other inconsistencies of the CFT are based on the fact that, in spite of a mainly ionic bond, the splitting pattern is caused predominantly by the orbital interactions between the d-orbitals of the metal ion and the ligand-orbitals instead of by purely electrostatic interactions as is assumed in CFT. It was shown (Lebernegg et al. 2008-2010) that the crystal-field splitting depends on the overlap integrals as:

$$\Delta_{CF} = \bar{v}_{pd\sigma}\left(1.5S^2_{2s,3d;\sigma} + S^2_{2p,3d;\sigma} - 0.9S^2_{2p,3d;\pi}\right) + \bar{v}^2_{pd\sigma}\left(-\frac{2.25 \cdot S^2_{2s,3d;\sigma}}{\Delta H_{sd}} - \frac{S^2_{2p,3d;\sigma}}{\Delta H_{pd}} + \frac{0.81 \cdot S^2_{2p,3d;\pi}}{\Delta H_{pd}}\right) -$$

$$- \frac{\Delta H_{sd}}{4}S^2_{2s,3d;\sigma} - \frac{\Delta H_{pd}}{4}S^2_{2p,3d;\sigma} + \frac{\Delta H_{pd}}{4}S^2_{2p,3d;\pi}. \qquad (3.2)$$

where ΔH_{id} =$H_{ii} - H_{dd}$ is the energy difference between the atomic orbitals in a molecule and the overlap integrals are

$$S_{2s,3d;\sigma} = \frac{x^2 e^{-x}}{45\sqrt{6}}(x^3 + 3x^2 + 6x + 6), \qquad (3.3)$$

$$S_{2p,3d;\sigma} = \frac{-xe^{-x}}{45\sqrt{2}}(x^4 + x^3 - 9x^2 - 30x - 30) \qquad (3.4)$$

$$S_{2p,3d;\pi} = \frac{xe^{-x}}{15\sqrt{6}}(x^3 + 6x^2 + 15x + 15) \qquad (3.5)$$

and $\bar{v}_{pd\sigma}$ is the radial contribution of a potential corresponding $(pd\sigma)$-interactions that for octahedral symmetry in $[FeO_6]^{10-}$ complex is

$$\bar{v}_{pd\sigma} = \frac{-143.49}{R^2} . \qquad (3.6)$$

In equations (3.3-3.5) $x = \zeta R$, where R is the metal-ligand distance and ζ is the average of the orbital exponents of the two interacting Slater type orbitals. Therefore, the R-dependence of the crystal-field splitting is dominated by the exponential factor. Accordingly, the splitting between t_{2g} and e_g orbitals was approximated with respect to equation (3.2) (Fig. 3.3 c). This approximation fits the numerical values over a broad range of distances.

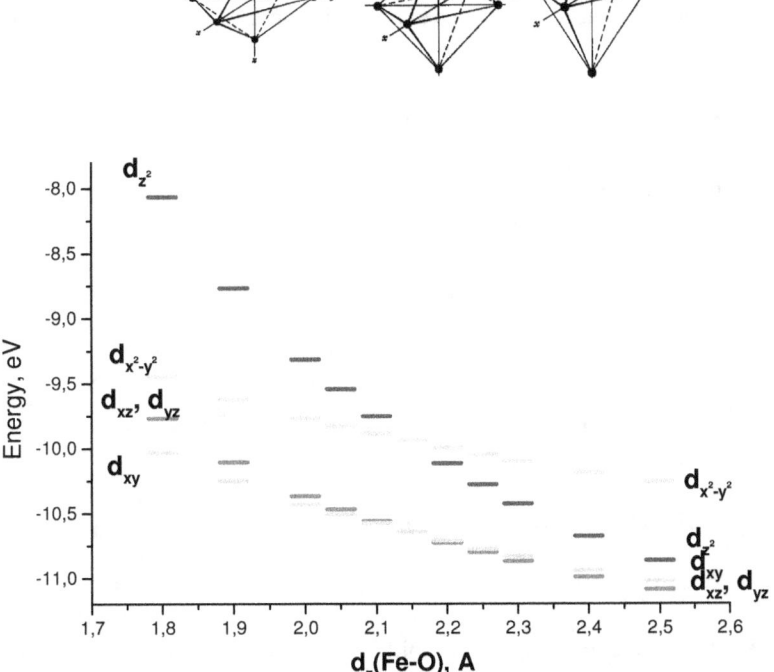

Figure 3.4 Splitting pattern of the $3d$ orbitals in tetragonally distorted octahedral sites with Fe-O distances of 2.15 Å in the xy plane and variable Fe–O distances (from 1.8 to 2.5 Å) along the z axis.

If the two oxygens on the z axis move towards the central Fe^{2+} or/and the four oxygen atoms in the xy plane move away, the symmetry is reduced from O_h to D_{4h}. As a consequence, the degenerate t_{2g} orbitals are separated into two sublevels with the d_{xy} orbital becoming more stable than the d_{xz} and d_{yz} orbitals, and the e_g orbitals are also split, $d_{x^2-y^2}$ becoming more stable than the d_{z^2} orbital. The single spin-down electron will occupy the d_{xy} orbital in which Coulomb repulsion will be minimized.

An inverse situation occurs if the two oxygen ions move away along the z axis or/and the four oxygen atoms in the xy plane move closer to the Fe^{2+} cation: d_{xy} and $d_{x^2-y^2}$ orbitals destabilize relative to d_{xz}, d_{yz} and d_{z^2} orbitals, respectively. The resulting octahedra with Fe–O distances of 2.15 Å in xy plane and various Fe–O distances along the z axis with corresponding calculated energies of the $3d$ orbitals are shown in Fig. 3.4. One of the most obvious results of this calculation is that both the energy separations between the d-orbitals and the destabilization of the orbital energies increase with decreasing distances. In addition, the d_{z^2} molecular orbital exhibits admixtures from the Fe(4s)-orbital increasing with decreasing Fe–O distances, up to 0.2 for d_z(Fe–O)=1.9 Å. Therefore, in equation (3.2) the contribution of the 4s-orbitals has to be included in the case of a tetragonally compressed octahedron. Furthermore, the splitting between the d_{z^2} and d_{xy} orbitals can be approximated with a high degree of accuracy by the power function $\sim R^{-n}$ yielding $n=6.58$ in the region of R typical for the Fe–O bond length (Fig. 3.5) which is larger than $n=5$ predicted in the frame of CFT. This deviation is due to the radial exponential dependence of the overlap integrals

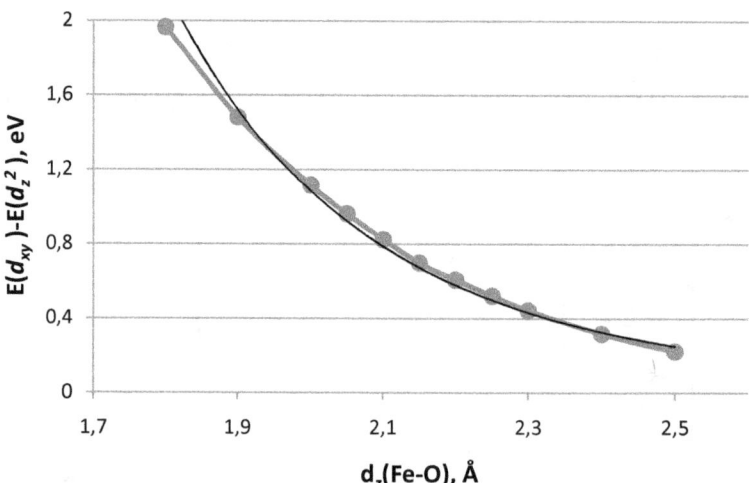

Figure 3.5 The calculated $d_{xy} - d_{z^2}$ splitting (dots) for a tetragonally distorted octahedral site with Fe–O distances of 2.15 Å in the equatorial plane and variable Fe–O distances (from 1.8 to 2.5 Å) along the z axis. A best-fit by the power function (thin black line), $\Delta_{CF} = \frac{103.8}{R^{6.58}}$.

(3.3-3.5) which are not included in the CFT.

In the next step, trigonal distortions of the octahedron will be discussed. If the three oxygen anions of the triangle below and above the central iron cation are displaced by equal distances along the trigonal axis, the symmetry is reduced from O_h to D_{3d}. The resulting trigonally elongated and compressed octahedron with corresponding calculated energies of the $3d$ orbitals are shown in Fig. 3.6. Due to the trigonal distortion, the degenerate t_{2g} orbitals of the ideal octahedron split into orbitals with symmetry a_{1g} and e'_g, whereas the upper e_g- orbitals remain degenerate (Fig. 3.6). It should further be noted that in the case of the trigonal elongation one of the two lowest spin-down e'_g orbitals and one of the two upper spin-down e_g orbitals have predominantly d_{xy} and d_{xz} – character, whereas the other two MOs have predominantly d_{yz} and $d_{x^2-y^2}$ – character, respectively, while the a_{1g} MOs has pure d_{z^2} character (Table 3.1).

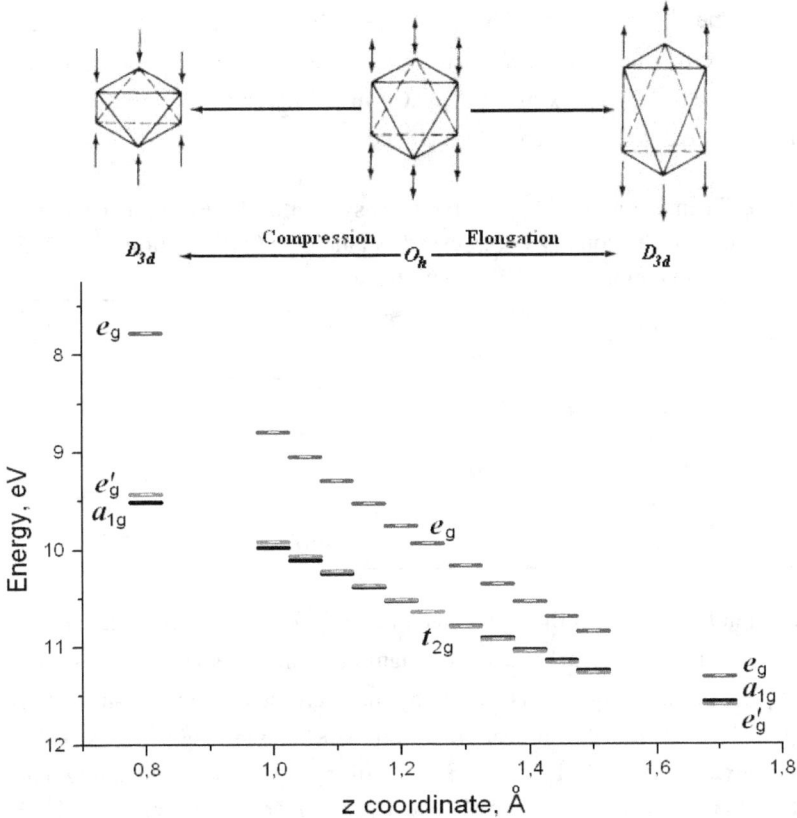

Figure 3.6 Splitting pattern of the $3d$ orbitals of Fe^{2+} ion in trigonally distorted octahedral site with variable z coordinate (from 0.8 to 1.7 Å) and constant x and y coordinates of 1.2413 Å for O anions.

Table 3.1 Eigenvectors in the principal axes system of the EFG for the spin-down Fe(3d)-orbitals in a trigonally elongated octahedron (contributions of the 3d atomic orbitals to the 3d-like molecular orbitals) with $z=1.5$ Å.

	$\varepsilon_0(\pi^*, e_g')$	$\varepsilon_1(\pi^*, e_g')$	$\varepsilon_2(\pi^*, a_{1g})$	$\varepsilon_3(\sigma^*, e_g)$	$\varepsilon_4(\sigma^*, e_g)$
ε_i, eV	11.265	11.265	11.243	10.840	10.840
d_{xy}	-0.857	0.001	0.000	0.000	-0.443
d_{yz}	0.000	0.472	0.000	0.810	0.000
d_{z^2}	0.000	0.000	-0.978	0.000	0.000
d_{xz}	-0.472	0.000	0.000	0.000	0.810
$d_{x^2-y^2}$	0.001	0.857	0.000	-0.443	0.000

The contributions do not change qualitatively in the case of the trigonal compression because the symmetry is preserved. The lowest a_{1g} orbital has $d_{z^2}-$ character, whereas the e_g' and e_g orbitals are mixtures of the d_{xy} and d_{xz}, d_{yz} and $d_{x^2-y^2}$ atomic orbitals which remain degenerate (Table 3.2). This mixing of the AOs in the e_g' and e_g orbitals is due to the axial crystal-field in the trigonal geometry (Gerloch and Slade 1973).

Table 3.2 Eigenvectors in the principal axes system of the EFG for the spin-down 3d-orbitals of Fe^{2+} in a trigonally compressed octahedron (contributions of the 3d atomic orbitals to the 3d-like molecular orbitals) with $z=0.8$ Å.

	$\varepsilon_0(\pi^*, a_{1g})$	$\varepsilon_1(\pi^*, e_g')$	$\varepsilon_2(\pi^*, e_g')$	$\varepsilon_3(\sigma^*, e_g)$	$\varepsilon_3(\sigma^*, e_g)$
ε_i, eV	9.514	9.429	9.429	7.781	7.781
d_{xy}	0.000	0.651	0.001	0.000	-0.722
d_{yz}	0.000	0.002	-0.740	0.632	0.000
d_{z^2}	-0.982	0.000	0.000	0.000	0.000
d_{xz}	0.000	0.740	0.002	0.000	0.632
$d_{x^2-y^2}$	0.000	0.001	-0.651	-0.722	0.000

It should be noted that the splitting between the a_{1g} and e_g' orbitals, reflecting the deviation from ideal O_h symmetry, is an order of magnitude smaller than the e_g'–e_g splitting. The increasing e_g'–e_g splitting (Fig. 3.7) is caused by the decrease of the metal-ligand distances and can be approximated by the power function $\sim R^{-n}$ yielding $n=2.51$ (Fig. 3.7) that considerably deviates from 7.71 and 6.58 for uniform expansion and tetragonal distortion, respectively. This deviation can be explained by considering the crystal field Hamiltonian based on the point charge model

$$H_{CF} \sim \sum_{lm} \frac{4\pi}{2l+1} \frac{r_<^l}{r_>^{l+1}} Y_{lm}(\hat{r}) Y_{lm}^*(\hat{R}_K). \qquad (3.7)$$

Due to the properties of the Gaunt integrals l is restricted to 0, 2 and 4. Next, terms corresponding to $l=0$ uniformly destabilize the d orbitals, terms with $l=2$ vanish for the ideal

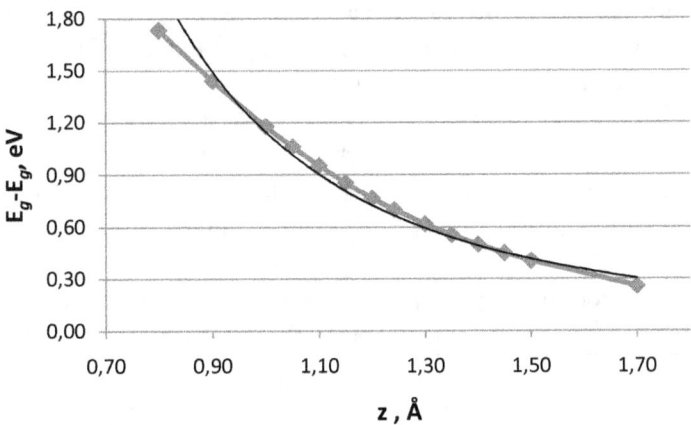

Figure 3.7 The calculated e_g'–e_g splitting (dots) for a trigonally distorted octahedral site with a best-fit by the power function (thin black line), $\Delta_{CF} = \frac{1,146}{R^{2.51}}$. Fe–O distances vary from 1.93 to 2.44 Å.

octahedron, so that only terms with l=4 remain giving rise to equation (3.1). For distorted octahedra terms with l= 2 do not vanish anymore leading to terms with lower power of $1/R$.

Another important class of distortions of the Fe-octahedron is angular distortions, e.g. the two opposite angles between the Fe–O bonds in the equatorial xy plane become smaller and,

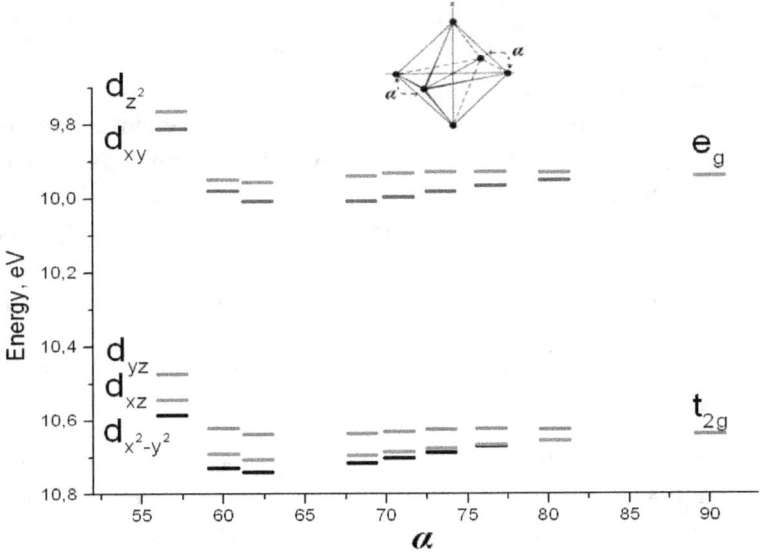

Figure 3.8 Splitting pattern of the Fe($3d$)- orbitals in angular distorted octahedral sites with variable opposite O–Fe–O angles in the equatorial xy plane and Fe–O distances of 2.15 Å.

consequently, the other two opposite angles between the Fe–O bonds in the xy plane become larger. Under this distortions the O_h symmetry is reduced to D_{2h} symmetry. In the calculations, the values of the two opposite O–Fe–O angles in the equatorial plane are varied from $57°$ to $90°$. The calculated energies of the $3d$ orbitals are shown in Fig. 3.8. The contributions of the $3d$ atomic orbitals to the $3d$-like molecular orbitals for an O–Fe–O angle of $60°$ are given in Table 3.3. The lowest $3d$ orbital has predominantly $d_{x^2-y^2}$ character with some contribution from the d_{z^2} orbital. In turn, the highest $3d$ orbital has predominantly d_{z^2} character with some contribution from the lowest $d_{x^2-y^2}$ orbital.

Table 3.3 Eigenvectors in the principal coordination system of the EFG for the spin-down Fe($3d$)-orbitals in the angular distorted octahedron with $\alpha=60°$.

	$\varepsilon_0\left(\pi^*,d_{x^2-y^2}\right)$	$\varepsilon_1\left(\pi^*,d_{x^2z}\right)$	$\varepsilon_2(\pi^*,d_{xz})$	$\varepsilon_3\left(\sigma^*,d_{xy}\right)$	$\varepsilon_4\left(\sigma^*,d_{z^2}\right)$
ε_i, eV	10.730	10.691	10.621	9.980	9.949
d_{xy}	0.000	0.000	0.000	-0.909	0.000
d_{yz}	0.000	0.000	-0.980	0.000	0.000
d_{z}^{2}	0.252	0.000	0.000	0.000	0.909
d_{xz}	0.000	0.982	0.000	0.000	0.000
$d_{x^2-y^2}$	-0.959	0.000	0.000	0.000	0.249

As a result of this angular distortion, the $d_{x^2-y^2}$ orbital which is oriented between the oxygen ions divides equatorial O–Fe–O angles in half becoming more stable. The spin-down electron will occupy this stabilised $d_{x^2-y^2}$ orbital and, therefore, the Coulomb repulsion of the single electron in this orbital by the oxygen ions is balanced. Consequently, the lobe of the d_{xy} orbital comes closer to the oxygen ions than d_{yz} and d_{xz} orbitals, and therefore it is more destabilized. However, the d_{xy} orbital is less destabilized than the d_{z^2} orbital which has lobe directed along the z-axis closer to the ligands than other $3d$-orbitals. In turn, as a result of angular distortion, the four oxygen ions in the equatorial xy plane move closer to the y axis and, simultaneously, farther from the x axis. Therefore, the d_{yz} orbital is destabilized, while the d_{xz} orbital is stabilized. Next, decreasing the angle below $62°$ leads to a subsequent destabilization of all the $3d$ orbitals. The overlap between the 2s and 2p orbitals of the approaching oxygens becomes significant, so that the destabilization is caused by the contribution of the Coulomb repulsion between the ligands. In addition, for a cubic or ideal triangular dodecahedral crystal-field of $(Fe^{2+}O_8)^{-14}$ with Fe–O distances of 2.292 Å, the calculated energy separation between the spin-down low-lying e_g orbitals and high-lying t_{2g} orbitals is 2767 cm^{-1}.

In summary, it can be concluded that the point charge model for describing the splitting pattern within the frame of CFT is inappropriate, whereas the model based on the overlap integrals has correct physical basis for the description of the crystal-field splitting in ideal and distorted polyhedra. Therefore, the investigation of crystal field splitting patterns has to be extended to distorted systems. Furthermore, despite the fact that the understanding of the physical and chemical properties of materials requires the knowledge about the splitting patterns, AO contributions, overlap integrals and other electronic properties for variable types and degree of distortions in the octahedral, tetrahedral and dodecahedral complexes, their systematic investigations are not existent in a sufficient range. The investigation for the tetragonal, trigonal and angular distortions of various degrees for the octahedral complex describing above is the first step in this way.

3.2 Mössbauer Spectroscopy

Another important experimental method yielding highly accurate results that can be compared with calculated data is *Mössbauer spectroscopy* that is based on the recoilless emission and resonant absorption of γ-rays by specific nuclei in solids (*Mössbauer effect*) (Mössbauer 1958).

Resonant absorption of γ-quants by free nuclei is not possible under normal conditions, because the recoil energy E_R of the γ quant is large compared with the line width Γ of nuclear transitions. However, if the emitting and absorbing nuclei are rigidly bound in a solid, the recoil energy transferred to the whole solid yielding $E_R \rightarrow 0$. Thus, both the recoil-free resonant emission and absorption become possible.

Since a nucleus is not rigidly bound, there is a certain probability that a nucleus excites phonons (lattice vibrations) at a finite temperature in a lattice and these phonon excitations lead to the loss of energy. Therefore, we are interested in the zero-phonon transitions. The *Debye-Waller factor* (recoil-free fraction, fraction of zero-phonon transitions, Lamb-Mössbauer factor) f_{DW}, which is a measure of the probability of recoil-free resonance (elastic scattering), is defined as:

$$f_{DW} = \exp\left(-\frac{4\pi^2 \langle x^2 \rangle}{\lambda^2}\right) = \exp\left(-\frac{E_\gamma^2 \langle x^2 \rangle}{\hbar^2 c^2}\right). \tag{3.8}$$

In equation (3.8), $\langle x^2 \rangle$ is the mean square vibrational amplitude of the emitting nucleus in the direction of the γ-ray, λ is the wavelength of the γ-quant, $\hbar$ is Planck's constant and $E_\gamma = \frac{2\pi\hbar c}{\lambda}$ is the recoil-free energy of the γ-quant. The recoil-free fraction depends on a nuclear property, *viz.* E_γ, and a crystal property, *viz.* $\langle x^2 \rangle$. The Debye-Waller factor f_{DW} becomes larger with decreasing E_γ. The mean square vibrational amplitude $\langle x^2 \rangle$ depends on the properties and the vibrational density of states (phonon spectrum) of the crystal lattice. The Debye model assumes that the angular frequency ω of a mode in a lattice is proportional to its wave-vector k and phase-velocity (sound speed) v_{pv} ($\omega = v_{pv}k$) and for low temperatures provides a good approximation for $\langle x^2 \rangle$ and equation (3.8) can be written as:

$$f_{DW}(T) = \exp\left(-\frac{3(\hbar k_\gamma)^2}{4Mk_B\Theta_D}\left[1 + \frac{2\pi^2}{3}\left(\frac{T}{\Theta_D}\right)^2\right]\right), \tag{3.9}$$

where k_B is the Boltzmann constant, Θ_D is the Debye temperature above which all normal modes of vibrations in a crystal can to be excited so that the crystal behaves classically, below this temperature the modes of lattice vibrations begin to be "frozen out" requiring quantum treatment of a crystal (Ashcroft and Merin 1976). Θ_D renormalized to the masses is a convenient parameter to characterize the strength of bonds between the atoms in the solid. In the case of the 14.413 keV transition in ^{57}Fe, f_{DW} is 0.91 at T=0 K (Gütlich et al. 1978). With decreasing of $\langle x^2 \rangle$, f_{DW} distinctly increases. Because $\langle x^2 \rangle$ usually decreases with falling temperature, f_{DW} increases with decreasing temperature; this is one reason why low-temperature measurements are very important in Mössbauer spectroscopy (Amthauer et al. 2004).

The strength of Mössbauer spectroscopy has two reasons. First, the most common and suitable Mössbauer active element is iron belonging to the five most abundant elements of the Earth. Accordingly, many important minerals contain iron, so that valuable information may be obtained from this method. It should be noted that mainly Mössbauer spectroscopy of ^{57}Fe containing minerals will be described due to the objects of our investigations. Second, the detection of very small energy differences is possible due to the high resolution and accuracy of the method. The natural line width Γ of the γ-ray following from the Heisenberg uncertainty principle is

$$\Gamma\tau = \hbar, \qquad (3.10)$$

with Planck's constant $\hbar$. Accordingly, the mean lifetime τ of $1.43 \cdot 10^{-7}$ s of the first excited state of ^{57}Fe yields a line width Γ of $4.55 \cdot 10^{-9}$ eV, which is extremely small compared to the energy, E_γ, of the corresponding γ-quant of $14.413 \cdot 10^3$ eV. This small line width enables the detection of extremely small relative changes of the γ-radiation energy $\Gamma/E_\gamma \sim 10^{-12} - 10^{-14}$. These relative changes are in the energy range of *hyperfine interactions*, i.e. interactions of the nuclear moments with electric and magnetic fields arising from the environment of the nucleus (Amthauer et al. 2004). These hyperfine interactions can shift and split nuclear energy levels due to the electric monopole, magnetic dipole or electric quadrupole interactions providing the physical, crystallographic and chemical information about the absorber. Therefore, it is important to understand the connection between hyperfine parameters and electronic structure.

The *electric monopole interaction* is the electrostatic interaction between the positive charge distribution of a nucleus of finite size and the negative electronic charge distribution with nonzero probability amplitude at the site of this nucleus (s-electrons in the nonrelativistic case). Since the nulear radii are different for the ground and excited states the monopole interaction generally causes a shift δ of the nuclear levels, the so-called *isomer shift* (IS) or *chemical shift*. The temperature-independent part δ_0 of the isomer shift, caused by differences of the averaged nuclear radii in the ground and excited state and by differences in the electronic and crystallographic environment of the source and the absorber, can be written as

$$\delta_0 = \frac{4\pi}{5} Ze \frac{\delta R}{R} R^2 [\rho_A(0) - \rho_S(0)] = \alpha[\rho_A(0) - \rho_S(0)]. \quad (3.11)$$

In equation (3.11), Ze is the nuclear charge, $\delta R = R_e - R_g$ (average nuclear radii in the excited, R_e, and ground, R_g, states), $\rho_A(0)$ and $\rho_S(0)$ are the electron densities at the nucleus at the absorber and the source, respectively. Therefore, the isomer shift δ_0 depends on two factors. The isomer shift calibration constant $\alpha = \frac{4\pi}{5} Ze \frac{\delta R}{R} R^2$ contains only nuclear parameters and is constant for a particular nuclear transition in a Mössbauer isotope. The discussion about the proper value of $\alpha(^{57}Fe)$ for the transition from the ground state with $I=1/2$ to the first excited state ($I = 3/2$) is still continuing but most of the determinations based on density functional methods cover (relativistic) values in the range of $(-0.27\pm0.04)a_0^3$ mm·s^{-1} (Grodzicki and Lebernegg, 2010). A negative value as in ^{57}Fe means that the average radius of the excited state with $I = 3/2$ is smaller than the radius of the ground state ($I = 1/2$). The second factor describes the difference between the electronic charge densities at the source and absorber nuclei which depend on the chemical bonding, spin state, oxidation and coordination number of the Mössbauer atom. In general, source and corresponding electronic charge density are standard, whereas the absorber is the sample under consideration. Hence, for the same source the isomer shift depends only on the electron density at the absorber nuclei and assuming α to be known, the measured isomer shift provides information about the electronic density at the absorber.

The following discussion will be restricted mainly on the two most common oxidation states, viz. high-spin (hs) Fe^{2+}, $3d^6$, and hs-Fe^{3+}, $3d^5$, because enough experimental data exist for establishing empirical correlations with a well-founded data basis. A selection of such empirical correlations is listed below:

(i) δ(hs-Fe^{2+}) is always considerably larger than δ (hs-Fe^{3+}); this is exploited as a common finger print technique to distinguish hs-Fe^{2+} and hs-Fe^{3+};

(ii) δ (hs-Fe^{2+}) distinctly increases with the coordination number provided the ligands are the same, e.g. for Fe-O compounds: δ(hs-$^{[4]}Fe^{2+}$) $\approx$ 1.0 mm·s^{-1}, δ(hs-$^{[6]}Fe^{2+}$) $\approx$ 1.25 mm·s^{-1}, δ(hs-$^{[8]}Fe^{2+}$) $\approx$ 1.4 mm·s^{-1};

(iii) δ(hs-Fe^{2+}) in $Fe^{2+}X_6$-compounds increases with the electronegativity of the ligands, e.g. δ(X=S) is reduced by about 0.2 - 0.3 mm·s^{-1} compared with δ(X=O), and in the hexahalides δ decreases in going from F to I. On the other hand, variations of δ(hs-Fe^{3+}) with regard to coordination number and ligands are much less pronounced and not so uniform;

(iv)   generally, δ is virtually independent of local distortions of the coordination polyhedron of Fe as far as the average distances remain approximately unchanged.

The origin of these trends can be traced back to the electronic structure. Firstly, from experimentally determined geometries it is known that Fe–O distances in hs-Fe^{2+} compounds are about 0.2 – 0.3 Å larger than in the analogous hs-Fe^{3+} compounds, for the reason that there is an additional electron in a (weakly) antibonding $\pi_\downarrow^*$-orbital. Secondly, it was demonstrated (Grodzicki and Lebernegg 2010) that the dominating quantities determining the isomer shift are the (average) distance between Fe and the ligands of the first coordination sphere and the oxidation state. Especially, with decreasing oxidation state the 4s-electrons (and to some extent the 3s-electrons) are more effectively screened by the 3d-electrons so that these orbitals become more diffuse. Consequently, the amplitudes of these s-orbitals at the nucleus are reduced and the isomer shift becomes larger though this decrease of $\rho(0)$ is distinctly smaller than expected from calculations on free ions. Indeed, the variation of the effective charge of iron is much less pronounced due to the increasing covalency of Fe in higher oxidation states. Furthermore, the increase of the isomer shift δ(hs-Fe^{2+}) with the coordination number, cf. (ii) above, is predominantly the result of the increasing Fe-ligand

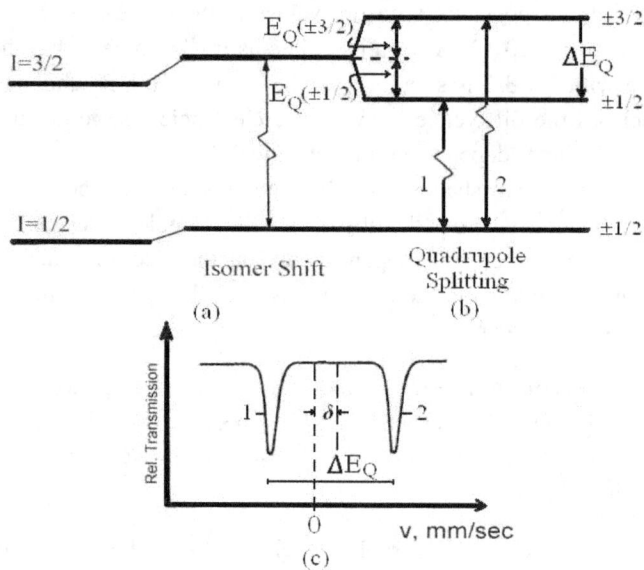

Figure 3.9 Schematic representations for:

(a)   the shift of nuclear level due to Coulomb interaction between the nuclear and electronic charge distributions;

(b)   the electric quadrupole interaction which splits the first excited state of the nucleus (I=3/2) in two sublevels by ΔE_Q;

(c)   the resultant Mössbauer spectrum.

distance when going from fourfold to eightfold coordinated compounds. This is easily demonstrated by model calculations on analogous systems with different coordination numbers but equal distances.

The second measurable quantity supplying information about the electronic structure of the environment of a Mössbauer nucleus is the quadrupole splitting. The quadrupole splitting, ΔE_Q, is an important parameter in the application of the Mössbauer effect in solid state physics, chemistry and crystallography, because ΔE_Q provides information on the symmetry, degree of distortion and coordination of the crystallographic site, number of non-equivalent positions and electronic configuration.

Generally, the electric quadrupole interaction splits a nuclear state with the spin quantum number I into $\frac{2I+1}{2}$ sublevels. Next, if the nuclear charge distribution deviates from spherical symmetry (nuclear spin $I>1/2$), the nucleus has a quadrupole moment. The values of Q are positive for an elongated positive charge distribution of a nucleus (cigar) and negative for a flattened nucleus (disc). The nuclear ground state of Fe has a spin of 1/2 and, correspondingly, no quadrupole moment. The first excited state of the Fe nucleus has a spin of 3/2 and more recent determinations show convergence towards a positive value of $Q=0.15\pm0.02$ barn (Ray and Das 1977; Lauer et al. 1979; Dufek et al. 1995; Blaha et al. 2000). Since ^{57}Fe has a non-zero quadrupole moment in the first excited state (I=3/2), the quadrupole interaction removes degeneracy and splits the excited state into two sublevels while the ground state is not split due to the vanishing quadrupole moment (Q=0). Therefore, only two transitions from the ground state to the split excited state are possible. Hence, the corresponding Mössbauer spectrum (Fig. 3.9) is a two line spectrum where quadrupole interaction separates two resonant absorption lines by the value of quadrupole splitting, ΔE_Q. The centroid of the two absorption lines relative to the zero velocity of the source corresponds to the isomer shift, δ.

Since, the quadrupole moment Q is constant for a certain Mössbauer isotope, differences in the quadrupole splitting can solely be attributed to changes of the EFG. If the electronic environment of a nucleus is asymmetric, i.e. deviates from cubic symmetry, the EFG arises from the surrounding charges. As a consequence, the electric quadrupole interaction takes place between a nuclear quadrupole moment Q and the EFG. In turn, the surrounding electric charges (electrons and lattice ions) determine the EFG-tensor as the second derivative of the potential V at the nucleus of all extranuclear charges. It is a 3x3 second-rank tensor

$$\nabla E = -\Delta V = -\begin{pmatrix} V_{xx} & V_{xy} & V_{xz} \\ V_{yx} & V_{yy} & V_{yz} \\ V_{zx} & V_{zy} & V_{zz} \end{pmatrix}. \tag{3.12}$$

The EFG-tensor is symmetrical, $V_{ij}=V_{ji}$, and according to Laplace's equation $\sum V_{ii} = 0$. Hence, the EFG-tensor can always be diagonalized by an appropriate choice of axes. In the principal-axes system of the EFG, all off-diagonal elements are zero. If the principal-axes system is chosen in a way that

$$|V_{ZZ}| \geq |V_{YY}| \geq |V_{XX}|, \tag{3.13}$$

it can be seen from

$$V_{XX} + V_{YY} + V_{ZZ} = 0 \tag{3.14}$$

that only two independent parameters remain. The parameters are the *electric field gradient*

$$V_{ZZ} = eq \tag{3.15}$$

and the *asymmetry parameter*

$$\eta = \frac{V_{XX}-V_{YY}}{V_{ZZ}}. \tag{3.16}$$

The latter describes the deviation of the EFG from axial symmetry and can take values $0 \leq \eta \leq 1$ according to equations (3.13) and (3.14).

If nuclear transition in ^{57}Fe from the ground state to the first excited state (1/2→3/2) takes place, the *quadrupole splitting* (ΔE_Q, or Δ, or QS) may be expressed as:

$$\Delta E_Q = \frac{1}{2} e Q V_{ZZ} \left(1 + \frac{\eta^2}{3}\right)^{1/2},$$

(3.17)

where e is the electronic charge, Q is the nuclear quadrupole moment, V_{ZZ} and η are the parameters of the electric field gradient.

Within the frame of a valence-electron-only molecular orbital (MO) method in combination with the LCAO approximation, the evaluation of the EFG tensor $V_{\alpha\beta}$ is based on dividing the charge density into the positive point charges Q_I^{ion} of the ionic cores and the charge distribution arising from the valence electrons (Grodzicki et al. 1987):

$$V_{\alpha\beta} = \Sigma_{I\neq0} Q_I^{ion} V_{\alpha\beta}(\vec{R}_{I0}) - e_0 \Sigma_{IJ,ij} P_{ij}^{IJ} \int \varphi_i^I(\vec{r} - \vec{R}_{I0}) V_{\alpha\beta}(\vec{r}) \varphi_j^J(\vec{r} - \vec{R}_{J0}) d^3r. \quad (3.18)$$

The Mössbauer atom is located at $\vec{R}_0$, $\vec{R}_{I0} = \vec{R}_I - \vec{R}_0$, $\varphi_i^I(\vec{r})$ is the (real) i^{th} atomic orbital centered at the Ith atom, P_{ij}^{IJ} is the bond-order matrix, and the tensor operator components $V_{\alpha\beta}(\vec{r})$ are given as

$$V_{\alpha\beta}(\vec{r}) = \frac{1 - \gamma(\vec{r})}{\vec{r}^3} \cdot \frac{3\vec{r}_\alpha \vec{r}_\beta - \vec{r}^2 \delta_{\alpha\beta}}{\vec{r}^2},$$

(3.19)

where the Sternheimer shielding function, $\gamma(r)$, describes the polarization of the core electrons of the Mössbauer atom by charges outside the core. With respect to a basis set of AO's the EFG tensor can be decomposed into three different contributions according to the location of the atomic orbitals in the matrix elements (3.18):

(1) <u>Valence contribution</u> with both atomic orbitals at the site of the Mössbauer atom:

$$V_{\alpha\beta}^{val} = -e_0 \Sigma_{ij} P_{ij}^{00} \int \varphi_i^0(\vec{r}) V_{\alpha\beta}(\vec{r}) \varphi_j^0(\vec{r}) d^3r.$$

(3.20)

The radial part of this one-center integral is proportional to $\langle r^{-3}\rangle(1 - R)$ where $R = \langle \gamma(r)r^{-3}\rangle/\langle r^{-3}\rangle$ is the Sternheimer shielding constant that accounts for the polarization of the inner closed shells by the valence electron distribution. The expectation value $\langle r^{-3}\rangle$ is derived from relativistic atomic calculations as a function of the valence shell occupation numbers. The shielding constant for ^{57}Fe is weakly dependent on the oxidation state of iron and covers values in the range between about 0.08 for higher oxidation states and 0.15 (low oxidation states).

(2) <u>Covalence contribution</u> with the atomic orbitals at the Mössbauer atom and a ligand site, respectively:

$$V_{\alpha\beta}^{cov} = -e_0 \Sigma_{J,ij} P_{ij}^{0J} \int \varphi_i^0(\vec{r}) V_{\alpha\beta}(\vec{r}) \varphi_j^J(\vec{r} - \vec{R}_{J0}) d^3r,$$

(3.21)

that describes the anisotropy of the electronic charge distribution in the iron-ligand bonds. In these two-center integrals $\gamma(r)$ appears explicitly. The numerical result for $\gamma(r)$ derived from

52

atomic self-consistent first-order perturbation calculations (Lauer et al. 1979) can be fitted with high accuracy by the expression

$$\gamma(r) = \gamma_\infty[f(\vec{r}) - f(0)] + \sum d_i exp[-c_i(\vec{r} - \vec{r}_i)^2], \qquad (3.22)$$

where $f(r) = \frac{1}{exp[-a(r-b)]+1}$, a, b, c_i, d_i, r_i are parameters determined by fitting the numerical results (Paulsen et al. 1994, Grodzicki et al. 1981), and γ_∞ having a value about -9 is the asymptotic value of $\gamma(r)$ for charges far outside the core. With this analytical representation of $\gamma(r)$ the radial part of the integrals (3.21) is calculated by numerical integration whereas the angular part can be evaluated analytically.

(3) <u>Ligand contribution</u> with both atomic orbitals at ligand atoms:

$$V_{\alpha\beta}^{lig} = \sum_{I\neq0} Q_I^{ion} V_{\alpha\beta}(\vec{R}_{I0}) - e_0 \sum_{IJ,ij} P_{ij}^{IJ} \int \varphi_i^I(\vec{r} - \vec{R}_{I0}) V_{\alpha\beta}(\vec{r}) \varphi_j^J(\vec{r} - \vec{R}_{J0}) d^3r, (3.23)$$

where the sum in the electronic part runs over all ligands, i.e. $I,J \neq 0$. In this case, the Sternheimer shielding function $\gamma(r)$ can be replaced with γ_∞. It should be emphasized that this definition is not identical with the meaning of the lattice term in crystal-field or point-charge approximations.

The detailed evaluation of all these integrals has been described in Grodzicki et al. (1987). In the principal axes system of the EFG tensor the EFG V_{ZZ} is simply the sum of the valence, covalence and ligand contributions. The usually dominating valence term is roughly proportional to the anisotropies

$$\Delta n_d = n_{x^2-y^2} + n_{xy} - n_{z^2} - \frac{n_{xz}+n_{yz}}{2}, \qquad (3.24)$$

$$\Delta n_p = \frac{n_x+n_y}{2} - n_z, \qquad (3.25)$$

of the valence d-shell and p-shell occupations n_i, respectively, of the Mössbauer atom. It is important to note that this proportionality between V_{ZZ} and the valence shell anisotropies only holds with regard to the principal axes system (Grodzicki and Amthauer 2000, Grodzicki and Lebernegg, 2010). Otherwise, erroneous conclusions are obtained. It is often assumed that for high-spin ferric iron the ligand contribution should be the largest one. However, several calculations have demonstrated that the valence contribution generally dominates even when the anisotropy should vanish according to crystal-field theory (Keutel et al. 1999, Grodzicki et al. 2000, Grodzicki et al. 2001, Lougear et al. 2000).

In the next step, the calculation of the electron density at the nucleus is described. Assuming core-valence separability – in accordance with the basic assumption of any valence-electron-only MO method – the total charge density at the Mössbauer nucleus is split into a sum of valence and core contributions:

$$\rho(0) = \rho_v(0) + \rho_c(0) = \sum_k n_k|\psi_k(0)|^2 + 2\sum_k|\psi_k(0)|^2. \qquad (3.26)$$

Representing the MOs $\psi_k(\vec{r})$ in LCAO approximation with atomic orbitals $\varphi_i^I(\vec{r} - \vec{R}_I)$, the valence contribution consists of the one-center term only:

$$\rho_v(0) = \sum_{IJ} \sum_{ij} P_{ij}^{IJ} \; \varphi_i^I(\vec{R}_{I0}) \varphi_j^J(\vec{R}_{J0}). \tag{3.27}$$

Among the various terms those with $I \neq 0$ or $J \neq 0$ have turned out to be negligible, and therefore the valence contribution consists of the one-center term only:

$$\rho_v(0) = \sum_{ij} P_{ij}^{00} \; \varphi_i^0(0) \varphi_j^0(0). \tag{3.28}$$

The evaluation of the core density $\rho_c(0)$ is somewhat more involved since the core orbitals of the free atom are not orthogonal to the MOs of the valence-electron-only MO calculation as required for applying equation (3.26). Moreover, after orthogonalizing the core orbitals to the MO basis, these are no longer orthogonal to each other, so that they have to be reorthogonalized. Accordingly, the final expression for the Schmidt orthogonalized core orbital $\psi_n^{(c)}(\vec{r})$ is obtained as

$$|\psi_n^{(c)}\rangle = N_n \left(1 - \sum_k |\psi_k\rangle\langle\psi_k| - \sum_{m=1}^{n-1} |\psi_m^{(c)}\rangle\langle\psi_m^{(c)}| \right) |\varphi_n^{(c)}\rangle = \tag{3.29}$$

$$= N_n \left(\varphi_n^{(c)}(\vec{r}) - \sum_{IJ} \sum_{ij} S_{ni}^{cI} p_{ij}^{IJ} \; \varphi_j^J(\vec{r} - \vec{R}_{J0}) - \sum_{m=1}^{n-1} S_{mn}^{cc} \psi_m^{(c)}(\vec{r}) \right) \tag{3.30}$$

where the following abbreviations have been used:

$$|\varphi_n^{(c)}\rangle = \varphi_n^{(c)}(\vec{r}) \qquad\qquad = n\text{-th core orbital of the free atom,}$$

$$p_{ij}^{IJ} = \sum_{k,occ} c_{ik}^I c_{jk}^J \qquad\qquad = \text{"reduced" bond-order matrix,}$$

$$S_{ni}^{cI} = \int \varphi_n^{(c)}(\vec{r}) \varphi_i^I(\vec{r} - \vec{R}_{I0}) d^3r \quad = \text{core-valence overlap matrix,}$$

$$S_{mn}^{cc} = \int \psi_m^{(c)}(\vec{r}) \varphi_n^{(c)}(\vec{r}) d^3r \qquad = \text{core-core overlap matrix} = -N_m \Delta_{mn} \text{ in good}$$

approximation with

$$\Delta_{mn} = \sum_{IJ,ij} S_{mi}^{cI} p_{ij}^{IJ} S_{jn}^{Jc} \text{ and } N_n = (1 - \Delta_{nn})^{-1/2} = \text{normalization constant.}$$

The amplitude of the completely orthogonalized core orbital at the nuclear site of the Mössbauer atom can finally be written as

$$\varphi_n^{(c)}(0) = N_n \left(\varphi_n^{(c)}(0) + \delta_n^{MO}(0) + \delta_n^{AO}(0) \right), \tag{3.31}$$

where

$$\delta_n^{MO}(0) = -\sum_{IJ} \sum_{ij} S_{mi}^{cI} p_{ij}^{IJ} \; \varphi_j^J(\vec{R}_{0J}), \tag{3.32}$$

$$\delta_n^{AO}(0) = +\sum_{m=1}^{n-1} N_m \psi_m^{(c)}(0) \Delta_{mn}. \tag{3.33}$$

Since the core orbital chosen as the first one in this procedure has to be orthogonalized only to the valence molecular orbitals, $\delta_1^{AO}(0)$ vanishes and the AO contribution $\delta_n^{AO}(0)$ is successively computed from the previously obtained $\psi_1^{(c)}(0), \psi_2^{(c)}(0), \dots, \psi_{n-1}^{(c)}(0)$.

The central problem of this procedure is to combine consistently the atomic data describing the core orbitals with the valence orbitals from the MO calculation. On one hand, the amplitudes $\psi_n^{(c)}(0)$ and $\psi_i^0(0)$ of the core and valence orbitals at the Mössbauer nucleus are required a very high degree of accuracy since the density differences $\Delta\rho(0)$ are small differences of large numbers, in most cases less than $5a_0^{-3}$ compared with total (relativistic) densities of about 15070 a_0^{-3} for ^{57}Fe, i.e. less than about 0.03% of the total value. On the other hand, the valence orbitals are represented usually with small basis sets and the core orbitals are not included, at all, in a valence-electron-only method. For these reasons, the following strategy has been pursued in calculating $\rho(0)$. The core orbitals of the free atom enter $\psi_n^{(c)}(0)$ in two different ways, namely through the amplitudes $\varphi_n^{(c)}(0)$ and through the core-valence overlap matrix S_{ni}^{cl}. Both quantities are calculated along different lines according to the required degree of accuracy:

(i) The amplitudes $\psi_n^{(c)}(0)$ and $\psi_i^{(0)}(0)$ of both the core and valence orbitals are determined at the nuclear radius by highly accurate, fully numerical (i.e. basis set free) relativistic atomic calculations for a certain number of valence shell configurations. In case of ^{57}Fe, e.g., these are $4s^n 3d^m$ with $n = 0,1,2$ and $m = 4,5,6,7$ except the configuration $4s^2 3d^7$. From these eleven values interpolation formulas are derived as a function of the valence shell occupation numbers n_s and n_d for each atomic core and valence orbital of the Mössbauer atom with nonvanishing amplitude at the nucleus:

$$\sqrt{4\pi}\varphi_{rel}(0) = a_0 + a_s n_s + a_d n_d + a_{ss} n_s^2 + a_{sd} n_s n_d + a_{dd} n_d^2. \quad (3.34)$$

It should be emphasized that the valence orbitals of the Mössbauer atom are included in this procedure because the size of the MO basis set, especially for large systems, is usually too small for supplying values of sufficient accuracy for $\psi_i^{(0)}(0)$ (Grodzicki et al., 2003, Geiger et al. 2003).

(ii) In a fully relativistic treatment the $p_{1/2}$-orbitals should be included, as well, since they have nonvanishing amplitudes at the nucleus and, thus, contribute to the electron density $\rho(0)$. It has been shown, however, that electron density differences are affected much less: even in the heavy main-group Mössbauer atoms Sn and I the $p_{1/2}$-electrons contribute less than $0.5a_0^{-3}$ (Grodzicki et al. 1987, Männing and Grodzicki 1986), and for Fe their contribution is still an order of magnitude smaller. Consequently, in discussing electron density differences, the contribution of the $p_{1/2}$-electrons is negligible with regard to the expected accuracy of the calculations.

(iii) The required accuracy for computing the core-valence overlap matrix S_{ni}^{cl} allows choosing the core orbitals $\varphi_n^{(c)}(\vec{r})$ as the non-relativistic Hartree-Fock orbitals of the neutral atom (Clementi and Roetti 1974). Additionally, care has to be taken of the phase factors that have to be compatible with those of the AO basis of the molecular calculation. Since for the latter $\varphi_n^{(c)}(r) > 0$ for $r \to \infty$, this phase factor has to be $(-1)^{n-l+1}$.

The advantage of this procedure is that relativistic values for $\rho(0)$ can be obtained without performing a fully relativistic all-electron molecular calculation. This is particularly important for the heavy main-group Mössbauer atoms Sn, Sb, Te and I, but even for Fe the ratio between relativistic and nonrelativistic $\rho(0)$ values is about 1.3 which has to be accounted for when deriving the calibration constant α from the correlation between measured isomer shifts and calculated $\rho(0)$ values. The value of α derived by the SCC-Xα method from such a correlation is -0.28 mm·s^{-1} and, thus, well within the range of (-0.27±0.04) mm·s^{-1} mentioned above.

Another representative example for the reliability of this approach constitutes the discussion about the proper value of α and of the related change of the nuclear radius $\Delta R/R$, respectively, of ^{119}Sn during the Mössbauer transition (Grodzicki and Lebernegg 2010). Up to the early 1980s considerable disagreement existed about the magnitude of $\Delta R/R$ with values ranging from $-2.5\cdot10^{-4}$ up to $+3.6\cdot10^{-4}$. Calculations during the years 1983 and 1984 on a series of 34 tin-containing compounds by the SCC-Xα method in combination with the procedure for evaluating $\rho(0)$ just described yielded $\Delta R/R = 1.61\cdot10^{-4}$ (Männing and Grodzicki 1986). This value has been criticized as too small on the basis of two other calculations of comparable sophistication. A Scattered-Wave-Xα calculation gave $\Delta R/R = 2.00\cdot10^{-4}$ (Winkler et al. 1987), whereas an even larger value of $2.20\cdot10^{-4}$ was derived by the Discrete-Variational-Xα method (Terra and Guenzburger 1989). Especially, the latter authors discuss in more detail the reason why the SCC-Xα result should be wrong. Amazingly, two years later the same authors published a new value by a refined calculation yielding $\Delta R/R = 1.58\cdot10^{-4}$ (Terra and Guenzburger 1991), now almost identical with the SCC-Xα result. In the meantime, from another computation a value of $1.57\cdot10^{-4}$ (Yanaga et al. 1990) had been obtained, and the most recent attempt employing the Hartree-Fock (HF) approximation supplemented with a second order perturbation (MP2) calculation comes to values of $1.52\cdot10^{-4}$ (HF) and $1.70\cdot10^{-4}$ (MP2) (Kurian and Filatov 2009). In summary, these various calculations seem to converge to a "consensus value" around $1.6\cdot10^{-4}$ virtually identical with the original SCC-Xα result.

3.3 Magnetic Interactions

Magnetic properties of matter have been fascinating people since antiquity (Thales of Miletus 634-546 BC). This particular interest has led humanity to investigate and apply the phenomenon of magnetism from nautical compass to computers and particle accelerators. The study of magnetic properties of materials on the basis of the theories of Faraday, Maxwell, Zeeman, Bohr, Stern and Gerlach, Heisenberg, Anderson, Goodenough, Kanamori et al. gives a good idea of the nature of the magnetic interactions (Emeleus 1977).

The idea that the interaction between adjacent magnetic dipoles distinguishes ferromagnetic materials from paramagnetic ones was proposed by Sir Alfred Ewing and Pierre Curie (Ginsberg 1971). Pierre-Ernest Weiss proposed a theory where the interaction between atomic dipoles is replaced by a single internal magnetic field. However, the comparison of the thermal energy kT_c (10^{-4}–10^{-2} eV for T_c=10–1000K), which destroys ferromagnetic ordering, with the resulting energy of magnetic interaction between two classical magnetic dipoles (10^{-7} eV) shows that this magnetic interaction cannot be the cause of magnetic ordering in a ferromagnet.

Later on it was shown that the origin of force between interacting atomic dipoles is electrostatic in nature. This was done by considering the H_2 molecule in Heitler-London (HL) approximation (Heitler and London 1927), where the Hamiltonian is

$$H = -\frac{\hbar^2}{2m}(\Delta_1 + \Delta_2) - Ze^2\left(\frac{1}{\vec{r}_{a1}} + \frac{1}{\vec{r}_{b1}} + \frac{1}{\vec{r}_{a2}} + \frac{1}{\vec{r}_{b2}}\right) + \frac{e^2}{\vec{r}_{12}} + \frac{Z^2e^2}{\vec{R}_{ab}}, \qquad (3.35)$$

where $\vec{r}_{a1} = \vec{r}_1 - \vec{R}_a$ and $\vec{R}_{ab} = \vec{R}_a - \vec{R}_b$ and subscripts a and b refer to the cores with charge Z and 1 and 2 to the electrons. Accordingly, this two-electron system can be at singlet and triplet states with energy separation of

$$E_{triplet} - E_{singlet} = \frac{2(J_{ab} - Q_{ab}S_{ab}^2)}{1 - S_{ab}^4}, \qquad (3.36)$$

where $S_{ab} = \int \varphi_a^*(\vec{r} - \vec{R}_a)\varphi_b(\vec{r} - \vec{R}_b)\,d\vec{r}$ is an overlap integral, Q is the coulomb integral

$$Q_{ab} = \int \varphi_a^*(\vec{r}_{a1})\varphi_b^*(\vec{r}_{b2})\left(-\frac{Ze^2}{\vec{r}_{a1}} - \frac{Ze^2}{\vec{r}_{b2}} + \frac{e^2}{\vec{r}_{12}} + \frac{Z^2e^2}{\vec{R}_{ab}}\right)\varphi_a(\vec{r}_{a1})\varphi_b(\vec{r}_{b2})\,d\vec{r}_1 d\vec{r}_2 \qquad (3.37)$$

and J is an "exchange integral"

$$J_{ab} = \int \varphi_a^*(\vec{r}_{a2})\varphi_b^*(\vec{r}_{b1})\left(-\frac{Ze^2}{\vec{r}_{a2}} - \frac{Ze^2}{\vec{r}_{b1}} + \frac{e^2}{\vec{r}_{12}} + \frac{Z^2e^2}{\vec{R}_{ab}}\right)\varphi_a(\vec{r}_{a1})\varphi_b(\vec{r}_{b2})\,d\vec{r}_1 d\vec{r}_2. \qquad (3.38)$$

The singlet-triplet separation (3.36) may be used to define an effective exchange between electrons. If $\hat{s}_1$ and $\hat{s}_2$ are the spin angular momentum operators of the two electrons then

$$\hat{s}_1 \cdot \hat{s}_2 = \frac{1}{2}\left(S(S+1) - \frac{3}{2}\right), \qquad (3.39)$$

where S=1 and $\hat{s}_1 \cdot \hat{s}_2 = \frac{1}{4}$ for the triplet state, and S=0 and $\hat{s}_1 \cdot \hat{s}_2 = -\frac{3}{4}$ for the singlet state. It is possible to introduce an effective Hamiltonian

$$H = -2J_{ab}\hat{s}_1 \cdot \hat{s}_2, \qquad (3.40)$$

for describing the magnetic excitation with

$$J_{ab} = -\frac{1}{2}\left(E_{triplet} - E_{singlet}\right) = -\frac{J_{ab} - Q_{ab}S_{ab}^2}{1 - S_{ab}^4}, \qquad (3.41)$$

so that we have a singlet ground state for $J_{ab} < 0$ and the ground state is a triplet if $J_{ab} > 0$.

It is important to emphasize that the exchange integral (3.38) includes two one-electron terms describing the Coulomb attraction between nuclei and electrons, and therefore (3.38) can be negative, whereas the "proper" exchange integral

$$J_{ab} = \frac{1}{2} \int \varphi_a^*(\vec{r}_{a2}) \varphi_b^*(\vec{r}_{b1}) \frac{e^2}{|\vec{r_1} - \vec{r_2}|} \varphi_a(\vec{r}_{a1}) \varphi_b(\vec{r}_{b2}) d\vec{r_1} d\vec{r_2} \qquad (3.42)$$

is always positive. Many early conclusions were based on the erroneous identification of J in (3.38) with J in (3.42), e.g., the conclusion has been drawn that magnetic coupling mediated by formally diamagnetic ligands must be positive

After this work an exchange interaction model was formulated (Heisenberg 1928), the famous Heisenberg Hamiltonian, which formally builds on the HL model. By extension of the two-spin Hamiltonian to a many-electron system with localized spins, the Heisenberg Hamiltonian is written as

$$H = -2 \sum_{i<j}^N J_{ij} \hat{s}_i \cdot \hat{s}_j. \qquad (3.43)$$

The significance of the Heisenberg Hamiltonian (3.43) expressed in the $\hat{s}_i \cdot \hat{s}_j$ form to the problems of magnetism has been realized (Van Vleck 1932). Moreover, if all the orbitals on the same center are assumed to be identical, then

$$H = -2 \sum_{a<b}^N J_{ab} \hat{S}_a \cdot \hat{S}_b, \qquad (3.44)$$

where

$$\hat{S}_a \cdot \hat{S}_b = \sum_i \hat{s}_{ai} \cdot \sum_j \hat{s}_{bj}. \qquad (3.45)$$

The summation in equation (3.44) is over pairs of centers a and b (atoms or ions) rather than pairs of electrons i and j. Equation (3.44) is the so-called *Heisenberg-Dirac-van Vleck* (HDVV) Hamiltonian and it is valid approximation for describing magnetic interactions in insulators and semiconductors with spins localized on each magnetic center.

Next, it should be noted that the *spin-coupling constant* J_{ab} decreases exponentially with increasing distance between the magnetic centers and this decreasing is due to the exponential decay of the wavefunctions $\varphi(\vec{r})$.

Due to the large distances (3–6 Å) between the magnetic centers in magnetic non-metallic systems any direct exchange mechanism is expected to be very small. With regard to the large distances between the interacting magnetic ions via diamagnetic ligands this interaction has been denoted as *superexchange*. Based on the assumption of ionic metal-ligand bonds, the exchange mechanism has been assigned to spin-polarization of the ligand electrons induced by the unpaired spins of the magnetic centers (Kramers 1934).

During the 1950's, however, EPR measurements supplied clear evidence that, in spite of predominantly ionic metal-ligand bonds, the magnetic orbitals must contain considerable admixtures from the ligand orbitals. Therefore, Anderson (1959) proposed an alternative exchange mechanism due to (weak) delocalization of the magnetic orbitals over the intervening diamagnetic ligands. Since delocalization leads to a decrease in kinetic energy, Anderson denoted this mechanism as kinetic exchange, and it is now generally acknowledged that, in most cases, this is the dominating mechanism for antiferromagnetic interaction in non-metallic systems. Accordingly, the coupling constant J consists of two contributions

$$J_{ab} = K_{ab} - 2\frac{t_{ab}^2}{U},$$

(3.46)

where the proper exchange integral

$$K_{ab} = \frac{1}{2} \int \varphi_a^*(\vec{r}_{a2}) \varphi_b^*(\vec{r}_{b1}) \frac{e^2}{|\vec{r}_1 - \vec{r}_2|} \varphi_a(\vec{r}_{a1}) \varphi_b(\vec{r}_{b2}) d\vec{r}_1 d\vec{r}_2$$

(3.47)

describes *potential exchange* which is always positive, whereas the second term

$$\frac{t_{ab}^2}{U} = \frac{[\int \varphi_a^*(\vec{r}_{a1}) H \varphi_b(\vec{r}_{b2}) d\vec{r}_1 d\vec{r}_2]^2}{\int \varphi_a^*(\vec{r}_{a1}) \varphi_b^*(\vec{r}_{b2}) \frac{e^2}{|\vec{r}_1 - \vec{r}_2|} \varphi_a(\vec{r}_{a1}) \varphi_b(\vec{r}_{b2}) d\vec{r}_1 d\vec{r}_2}$$

(3.48)

denoted as *kinetic exchange* represents the antiferromagnetic contribution. The quantity t_{ab} is a one-electron transfer or "hopping" integral describing the amount of delocalization and U is the coulomb repulsion between electrons. It was suggested that the strength of the kinetic exchange predominantly depends on the filling of the orbitals at the interacting magnetic centers and on the geometry (bridging angles and distances) of the metal-ligand-metal system. A satisfactory system of semi-empirical rules was developed (Goodenough 1958, Kanamori 1959). These *"Goodenough-Kanamori" rules* were formulated as (Anderson 1963):

1. When the two ions have lobes of magnetic orbitals pointing toward each other in such a way that the orbitals would have a reasonably large overlap integral, the exchange is antiferromagnetic. There are several subcases:

   a. when the lobes are d_{z^2} type orbitals in the octahedral case, particularly in the "180^0 position" in which these lobes point directly toward a ligand and each other, one obtains a large superexchange;

   b. when d_{xy} orbitals are in the 180^0 position to each other, so that they interact via p_π orbitals on the ligand, one again obtains antiferromagnetic superexchange;

   c. in a 90^0 ligand situation, when one ion has a d_{z^2} occupied and the other a d_{xy}, the p_π for one is the p_σ for the other and one expects a strong overlap and thus antiferromagnetic exchange.

2. When the orbitals are arranged in such a way that they are expected to be in contact but have no overlap integral, most notably a d_{z^2} and a d_{xy} in 180^0 position where the overlap is zero by symmetry, the rule gives ferromagnetic interaction.

In the case of a d^l binuclear complex, three different configurations give rise to low lying singlet state, while only one configuration is involved in the triplet state (Fig. 3.10). It was shown that the singlet state cannot be correctly described by a single configuration, because single determinant underestimate the relative stabilization of the singlet state relative to the triplet state (Noodleman et al. 1981, Ruiz et al. 1997) leading to an overestimation and qualitatively incorrect results of the spin-coupling constant (3.46). Therfore, the contribution of the two excited configurations to the lowest singlet state is important, and the neglecting these configurations is responsible for the poor description of the singlet state. In order to

59

solve this problem, a *broken-symmetry* (BS) solution was suggested (Noodleman et al. 1981). This method is based on a valence-bond description of the magnetic orbitals. In this approach, the low-spin state that corresponds to a "broken symmetry" wavefunction that is approximated by a linear combination of determinants

Figure 3.10 Singlet and triplet many-electron configurations.

$$\psi_{BS} = \Sigma_i\, a_i\, \varphi_i. \tag{3.49}$$

Using this wavefunction an expression for the (isotropic) Heisenberg coupling constant J is derived as

$$J = -\frac{E(S_{HS}) - E(S_{BS})}{S_{HS}^2 - S_{BS}^2}. \tag{3.50}$$

where $E(S)$ is the total energy for the spin state S. As mentioned above, positive values of J correspond to parallel or "ferromagnetic", negative values to antiparallel or antiferromagnetic coupling of the two spins $\vec{S}^A$ and $\vec{S}^B$. It is important to emphasize that both $E(S_{HS})$ and $E(S_{BS})$ are computed with respect to the same geometry so that both calculations merely differ by the relative orientation of two weakly coupled spins. The resulting extensive degree of "cancellation of unknown errors" enables the calculation of energy differences even below $10\ \text{cm}^{-1}$ (Grodzicki and Amthauer 2000, Grodzicki et al. 2009).

As mentioned above, the thermal energy destroys ferromagnetic ordering in the materials at Curie temperature T_C. In antiferromagnets, the critical temperature T_N above which magnetic ordering vanishes is known as Néel temperature. The type of the magnetic interaction in the materials can be also determined experimentally and there are several experimental methods. The measurements of the magnetic susceptibility as a function of the temperature is the most commonly used method. The value χ of the magnetic susceptibility is expressed as a function of temperature T accordingly to the Curie-Weiss law as

$$\frac{1}{\chi} = \frac{T - \theta}{C}, \tag{3.51}$$

where C is the Curie-Weiss constant determining the slope of the line given by equation (3.51) and Θ is the paramagnetic Curie temperature which can be determined as the intercept of the line with the temperature axis. Positive values for Θ indicate ferromagnetic interactions, while negative values indicate antiferromagnetic interactions (Fig. 3.11).

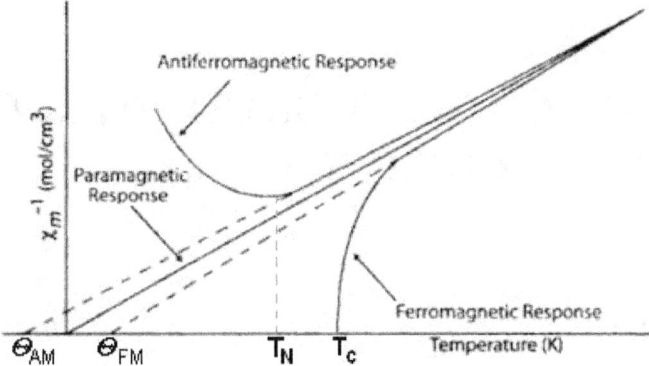

Figure 3.11 The inverse susceptibility versus temperature with the three major classes of magnetic behavior.

From these experiments, according to the mean-field theory (Weiss theory) the exchange coupling constant characterizing the interaction between magnetic atoms with spin S can be expressed via the paramagnetic Curie temperature as

$$J = \frac{3k_B\theta}{2S(S+1)},$$ (3.52)

Finally, it is important to emphasize an existence of the so-called *supersuperexchange* which represents significant contribution to the magnetic interaction between metal ions in complex M–O–O–M that frequently occurs in minerals. Though, there are attempts to approximate the supersuperexchange to the superexchange in order to describe it, the supersuperexchange is far less well understood, and therefore fundamental investigations of this magnetic interaction have to be done.

Chapter 4

ELECTRONIC AND MAGNETIC STRUCTURE OF ORTHOFERROSILITE

4.1 Introduction

Pyroxenes belonging to the general class of chain silicates are the most important group of rock-forming minerals constituting more than 20 vol% of the Earth's crust and upper mantle (Deer et al. 2001), and have also been found in minerals of extraterrestrial origin as in lunar and Martian rocks. In addition, during the last few years pyroxenes gained interest as a new class of multiferroic materials exhibiting both magnetic and ferroelectric properties at low temperatures (Jodlauk et al. 2007). Understanding the occurrence of ferroelectricity, however, requires the detailed knowledge of the magnetic structure. Recently, the electronic and magnetic structure of the ferrous end- member hedenbergite, $CaFe^{2+}Si_2O_6$; of the pyroxene quadrilateral $Mg_2Si_2O_6$-$CaMgSi_2O_6$-$CaFe^{2+}Si_2O_6$-$Fe^{2+}_2Si_2O_6$ has successfully been characterized by electronic structure calculations in the local spin density approximation (LSDA) (Grodzicki et al. 2009). As the next step of systematic investigation of iron-bearing silicates, the other iron end member of this quadrilateral, viz. the ferrous chain silicate orthoferrosilite, $Fe^{2+}_2Si_2O_6$, is investigated by the same methods in order to arrive at an improved understanding of the electronic and magnetic properties of the pyroxenes.

Under ambient conditions, orthoferrosilite crystallizes in the orthorhombic space group *Pbca* (Fig. 4.1) with unit cell parameters a = 18.418 Å, b = 9.078 Å and c = 5.2366 Å (Sueno et al. 1976). Fractional atomic coordinates for orthoferrosilite are presented in Table 4.1 with the corresponding coordinates of equivalent positions for the space group *Pbca*, No. 61 (Hahn 1987):

$$(1)\, x, y, z; \quad (2)\, \bar{x}+\tfrac{1}{2}, \bar{y}, z+\tfrac{1}{2}; \quad (3)\, \bar{x}, y+\tfrac{1}{2}, \bar{z}+\tfrac{1}{2}; \quad (4)\, x+\tfrac{1}{2}, \bar{y}+\tfrac{1}{2}, \bar{z};$$

$$(5)\, \bar{x}, \bar{y}, \bar{z}; \quad (6)\, x+\tfrac{1}{2}, y, \bar{z}+\tfrac{1}{2}; \quad (7)\, x, \bar{y}+\tfrac{1}{2}, z+\tfrac{1}{2}; \quad (8)\, \bar{x}+\tfrac{1}{2}, y+\tfrac{1}{2}, z.$$

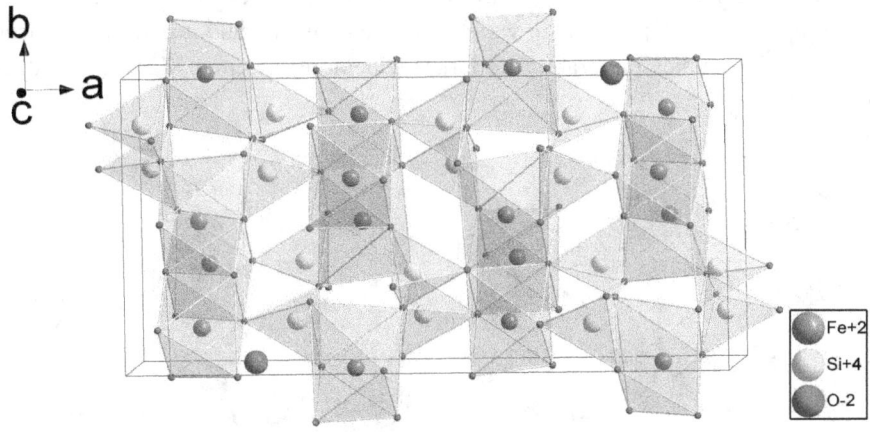

Figure 4.1 Unit-cell of orthoferrosilite.

The unit cell (Fig. 4.1) consisting of 80 atoms includes four similar zig-zag chains of edge-sharing Fe octahedra M1 (Fig. 4.2a) along the crystallographic c axis, which are enclosed by larger and strongly distorted M2 octahedra (Fig. 4.2b) resulting in a ribbon-like structure. In qualitative discussions, the M1 octahedron is assumed to be tetragonally compressed with approximate symmetry D_{4h} whereas the M2 octahedron is assumed to have pseudosymmetry C_{2v} (Burnham et. al. 1971, Sueno et. al. 1976, Yang and Ghose 1995, Hugh-Jones et. al. 1997). Within the chain two Fe(M1) atoms are separated by 3.14 Å corresponding to angles Fe(M1)-O1A-Fe(M1) of 95° and Fe(M1)-O1B-Fe(M1) of 93°. The shortest Fe(M1)–Fe(M2) distance is 3.00 Å with corresponding angles Fe(M1)-O1A-Fe(M2) of 93.6° and Fe(M1)-O1B-Fe(M2) of 89.6°.

Table 4.1 Fractional atomic coordinates for synthetic orthoferrosilite at 24°C (Sueno 1976)

Atom	No.	Oxidation	Site	x	y	z
Fe	1	+2	8 c	0.37573(3)	0.65415(7)	0.8746(1)
Fe	2	+2	8 c	0.37766(4)	0.48567(7)	0.3667(1)
Si	1	+4	8 c	0.27229(6)	0.3387(1)	0.0493(2)
Si	2	+4	8 c	0.47310(6)	0.33448(1)	0.7891(2)
O	1A	-2	8 c	0.1848(2)	0.3396(3)	0.0387(6)
O	2A	-2	8 c	0.3118(2)	0.4964(4)	0.0582(6)
O	3A	-2	8 c	0.3025(2)	0.2363(4)	0.8163(6)
O	1B	-2	8 c	0.5610(2)	0.3365(4)	0.7868(6)
O	2B	-2	8 c	0.4332(2)	0.4806(3)	0.6932(6)
O	3B	-2	8 c	0.4476(2)	0.2028(3)	0.5865(6)

Neighbouring ribbons are linked laterally by two crystallographically different chains of silicate tetrahedra parallel to the c axis. The two distinct chains are designated A and B (following the notation of Burnham, 1966). The B chain is the more kinked of the two, and the tetrahedra comprising this are larger and more distorted than those of the A chain. The tetrahedra of both chains are completely occupied by silicon. The shortest interchain Fe(M1)–

Fe(M1) and Fe(M1)–Fe(M2) distances are 5.52 and 4.84 Å, respectively. All natural iron-bearing orthopyroxenes (Opx) contain variable amounts of Mg, and it is well established that in this case Fe^{2+} occupies preferably the larger and more distorted M2 site (Goldman and Rossman 1979).

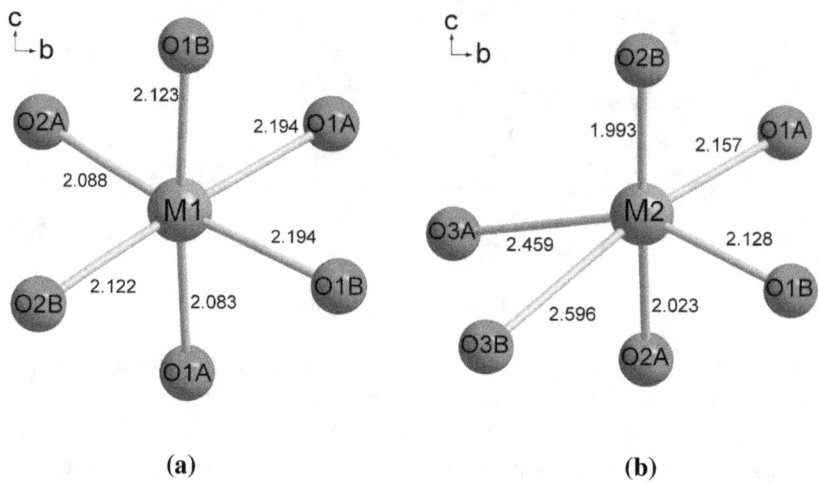

(a)   **(b)**

Figure 4.2 (100) projections of the M1 **(a)** and M2 **(b)** octahedra.

Orthoferrosilite and other members of the $Fe_2Si_2O_6$–$Mg_2Si_2O_6$ join were already subject to several experimental studies by optical spectroscopy (White and Keester 1966, Bancroft and Burns 1967, Runciman et al. 1973, Goldman and Rossman 1977, 1979, Steffen et al. 1988, Hiroi and Takeda 1992, Langer and Khomenko 1999, Taran and Langer 2001), by Mössbauer spectroscopy (Seifert 1983, Stanek and Hafner 1988, Steffen et al. 1988, Lin et al. 1993, Van Alboom et al. 1993, 1994) and by measurements of the magnetic properties (Shenoy et al. 1969, Wiedenmann and Regnard 1986, Regnard et al. 1986, Wiedenmann et al. 1986). The optical and near-to-mid-IR spectroscopic results can be summarized as yielding four bands due to spin-allowed d–d transitions at about 11000, 8500, 5000 and 2350 cm^{-1} with slightly increasing energies for decreasing iron content. With regard to the relation between symmetry and intensities and to the fact that Fe^{2+} preferably occupies the strongly distorted M2 site in natural Opx, the most intense bands at 11000 and 5000 cm^{-1} are assigned to transitions into the split e_g-like orbitals of Fe(M2). The mid-IR band at 2350 cm^{-1} is interpreted as a transition within the split t_{2g}-like orbitals of Fe(M2) whereas the weak band at 8500 cm^{-1}, observed only in iron-rich orthopyroxenes, is assigned to a transition into an e_g-like orbital of Fe(M1). The second t_{2g}-e_g transition at Fe(M1) is generally assumed to be hidden under the intense band at 11000 cm^{-1} but has more recently been identified with a weak shoulder at the low energy end of this band at 9300 cm^{-1} (Hiroi and Takeda 1992).

The Mössbauer spectrum of a synthetic orthoferrosilite at 77 K exhibits two clearly separated quadrupole doublets with quadrupole splittings ΔE_Q of 3.18 and 2.02 $mm \cdot s^{-1}$ (Steffen et al. 1988) that are assigned to Fe(M1) and Fe(M2), respectively. Experimentally derived asymmetry parameters η vary between 0.26 and 0.4 for Fe(M1), and between 0.16

and 0.35 for Fe(M2) (Wiedenmann et al.1986; Stanek and Hafner 1988; Van Alboom et al. 1993; Lin et al. 1993). With decreasing iron content ΔE_Q(M1) decreases to about 3.07 mm·s^{-1} whereas ΔE_Q(M2) increases up to 2.18 mm·s^{-1} (Lin et al. 1993; Van Alboom et al. 1993). Moreover, the quadrupole splitting of Fe^{2+}(M1) is strongly temperature dependent with ΔE_Q(M1) about 2.5 mm·s^{-1} at room temperature (RT) whereas ΔE_Q(M2) is almost constant up to RT. From the temperature dependence of ΔE_Q(M1) a splitting of 460 and 690 cm^{-1} for the t_{2g}-like orbitals of Fe^{2+}(M1) has been inferred (Van Alboom et al.1994).

Since iron in silicates is always in the high-spin state, orthoferrosilite exhibits, as virtually all iron-rich silicates, a magnetic phase transition from a disordered (paramagnetic) to an ordered antiferromagnetic state at low temperatures. The Néel temperature T_N for pure synthetic orthoferrosilites is about 38 K (Shenoy et al. 1969; Wiedenmann et al. 1986) and decreases with decreasing iron content down to 11 K for 0.758 iron apfu (Shenoy et al. 1969). From neutron diffraction patterns a collinear antiferromagnetic ordering is derived with alignment of the magnetic moments along the crystallographic b axis. The positive paramagnetic Curie temperature θ_p of +27.4 K (Wiedenmann et al. 1986) shows that the dominant magnetic interaction must be ferromagnetic. Accordingly, it has been concluded that the magnetic interaction within a ribbon is ferromagnetic whereas the interaction between neighbouring ribbons is antiferromagnetic and distinctly weaker with a ratio $|J_{inter}|/J_{intra} \approx 0.3$. Finally, a meta-magnetic transition is observed at an external field of 5.0 T (Wiedenmann et al. 1986).

In order to arrive at a more detailed understanding of the electronic and magnetic structure of Opx, the iron-end member orthoferrosilite is investigated by electronic structure calculations in the local spin-density approximation. The calculations are based on clusters of increasing size that model the local environment of the iron sites. The reliability of this theoretical approach will be assessed first by comparing the computed spectroscopic data with the experimental results. Afterwards, the magnetic structure is derived by calculating the Heisenberg coupling constants for the various intra- and inter-chain interactions. The results are analyzed and discussed by comparing the corresponding exchange pathways.

4.2 Electronic Structure of Orthoferrosilite

As mentioned in Section 2.3, the electronic structure calculations have been performed in the LSDA by the spin-polarized self consistent charge (SCC) $X\alpha$ method (Grodzicki 1980, 1985). The valence basis set consists of $2s$-, $2p$-orbitals for O and F (replacing incompletely coordinated oxygens, see below), $3s$-, $3p$-orbitals for Mg and Si, and $3d$-, $4s$-, $4p$-orbitals for Fe. The construction of suitable model clusters based on the experimentally determined geometry on a synthetic single crystal of orthoferrosilite at RT (Sueno et al. 1976) follows the experience from previous cluster calculations on minerals (Grodzicki et al. 2009 and references therein, Zherebetskyy et al. 2010).

In case of the M1 site of orthoferrosilite, the cations of the second coordination sphere comprise six (formally tetravalent) silicon atoms, and five divalent Fe cations. These are fourfold (Si) and sixfold (Fe) coordinated, respectively, by oxygen atoms. The smallest neutral cluster constructed along the rules described in Section 3.4 has the composition $[Fe^{2+}O_6\text{-}Si_6Fe_5O_2F_{20}]$. Adding stepwise further coordination polyhedra leads to a hierarchy of clusters of increasing size that may serve as a control for size convergence. The largest size-converged clusters used for the calculation of the spectroscopic data of iron at both crystallographic sites contain 143 atoms with the composition $Fe^{2+}O_6\text{-}Si_6Mg_5O_{30}\text{-}Si_{17}Mg_{17}O_5F_{30}$ for Fe at M1 and 136 atoms with the composition $Fe^{2+}O_6\text{-}Si_7Mg_3O_{28}\text{-}Si_{12}Mg_{22}O_3F_{54}$ for Fe at M2. These clusters represent the correct environment of the central iron atom within a sphere of 6.7 Å in the sense that all cations are included up to this distance and all anions are oxygen atoms. On the basis of these size-converged clusters additional electronic structure calculations were performed for a synthetic Mg-rich Opx, $(Mg_{0.75}Fe_{0.25})_2Si_2O_6$, with unit cell parameters a = 18.2747 Å, b = 8.8729 Å and c = 5.1988 Å (Yang and Ghose 1995), and for a natural mantle Opx with unit cell parameters a = 18.2429 Å, b = 8.8171 Å and c = 5.1894 Å and iron contents of 0.023 and 0.127 apfu at M1 and M2, respectively (Diego Gatta et al. 2007).

The first series of electronic structure calculations investigates the dependence on the cluster size of the spectroscopic data for Fe^{2+} at the M1 position (Zherebetskyy et al. 2010). The results given in Table 4.2 demonstrate that the two smallest clusters yield qualitatively wrong results, and size convergence is achieved for clusters with about 117 atoms. The theoretical quadrupole splitting for the size converged clusters is in quantitative agreement with the experimental values while the calculated d–d excitation energies into the σ-antibonding e_g-like orbitals are about 10% too large, in accordance with the experience from previous calculations on similar systems (Lottermoser et al. 2002, Geiger et al. 2003, Grodzicki et al. 2003, 2009).

Analogous calculations for Fe^{2+} at the M2 position again yield size convergence for clusters with about 110 atoms. The calculated spectroscopic data for a cluster of 136 atoms are compared with the available spectroscopic data in Table 4.3. The calculated quadrupole splitting is again in quantitative agreement with the experimental values, as is the calculated d–d excitation energy at about 5000 cm^{-1}. The highest excitation energy is obtained as 13615 cm^{-1} which is about 20% larger than the experimental value. The e_g-like orbitals are thus split by about 8500 cm^{-1} reflecting the strong tetragonal compression of the M2 octahedron with a ratio of $\bar{d}_{ax}/\bar{d}_{eq} = 0.86$ that is comparable with the tetragonal compression of the A-site of Fe^{2+} in vivianite resulting in a calculated splitting about 8000 cm^{-1} for the e_g-like orbitals, as well (Grodzicki and Amthauer 2000).

Table 4.2 Calculated and measured d–d excitation energies (in cm^{-1}) and quadrupole splittings ΔE_Q (in mm·s^{-1}) with η in brackets for orthoferrosilite clusters of increasing size around Fe^{2+}(M1)

N_{at}	$\varepsilon_1(\pi^*)$	$\varepsilon_2(\pi^*)$	$\varepsilon_3(\sigma^*)$	$\varepsilon_4(\sigma^*)$	$\Delta E_Q(\eta)$
7	186	282	5695	6638	-2.48 (0.79)
29	718	912	6904	8598	-2.71(0.14)
40	250	565	8034	9679	3.23(0.18)
63	307	468	8123	10429	3.29(0.74)
74	339	460	8010	10461	3.21(0.33)
89	323	476	8389	11010	3.22(0.31)
101	355	500	8623	11195	3.21(0.26)
117	363	508	8792	10744	3.18(0.18)
128	371	565	9139	10954	3.16(0.09)
143	379	589	9591	11422	3.13(0.16)
exp[a]	150	–	8500	10500[i]	3.181 (-)
exp[b]	–	–	8440	9314	– (–)
exp[c]	–	–	8640	10840[i]	– (–)
calc[d]	460	690	8450	9080	3.10[e] (0.26)
exp[f]	480	–	–	–	3.123 (–)
exp[g]	–	–	8300	10700[i]	– (–)
exp[h]	–	–	–	–	3.13 (0.4)

Measured ΔE_Q are all at LNT

[a] Steffen et al. (1988), $T = 77$ K

[b] Hiroi and Takeda (1992)

[c] Taran and Langer (2001)

[d] Van Alboom et al. (1994)

[e] Experimental: Van Alboom et al. (1993) with 0.27 Fe apfu, $T = 80$ K

[f] Lin et al. (1993) with 0.80 Fe apfu, T = 79 K; $\Delta E_Q = 3.131$ mm·s^{-1} at $T = 28$ K

[g] Average over five natural orthopyroxenes: Langer and Khomenko (1999)

[h] Wiedenmann et al. (1986), $T = 4.2$ K

[i] Incorrectly assigned to Fe^{2+}(M1), see text

Accordingly, with regard to the different average axial Fe–O distances, viz. 2.10 Å for M1 versus 2.01 Å for M2, it is rather unlikely that the highest excitation energies of these sites both should appear at virtually the same energy, but the calculated difference of about 2000 cm^{-1} rather supports the assignment of the highest d–d transition at M1 to a shoulder at 9300 cm^{-1} (Hiroi and Takeda 1992) which is also in accordance with a crystal-field theoretical analysis (Van Alboom et al. 1994).

Similarly, the calculated two low energy transitions at about 380 and 600 cm^{-1} are close to other estimates (Lin et al. 1993, Van Alboom et al. 1994) whereas a value of 150 cm^{-1} (Steffen et al. 1988) is definitely too small. The lowest d–d transition at the M2 site occurs at about 1000 cm^{-1} in accordance with the observed weak temperature dependence of ΔE_Q(M2) whereas a value of 354 cm^{-1} (Goldman and Rossman 1977) would yield a pronounced temperature dependence.

Table 4.3 Calculated and measured $d–d$ excitation energies (in cm^{-1}) and quadrupole splittings ΔE_Q (in mm·s^{-1}) with η in brackets for orthoferrosilite clusters of increasing size around Fe^{2+}(M2)

	$\varepsilon_1(\pi^*)$	$\varepsilon_2(\pi^*)$	$\varepsilon_3(\sigma^*)$	$\varepsilon_4(\sigma^*)$	$\Delta E_Q(\eta)$
7	822	992	3339	8066	1.33(0.22)
33	1129	1580	3831	10526	1.85(0.09)
39	992	1621	4033	11784	1.86(0.36)
48	1000	1613	4219	12058	1.86(0.42)
72	928	1435	4920	12696	2.12(0.27)
88	1121	1621	4904	13381	1.90(0.18)
99	1145	1580	4977	13542	1.95(0.12)
115	1185	1710	5017	13873	1.94(0.21)
136	1105	1581	5074	13615	2.01(0.23)
exp[a]	–	–	–	10500	2.021 (–)
exp[b]	–	–	5333	10914	– (–)
exp[c]	–	–	5460	10840	– (–)
calc[d]	960	2340	5390	11040	2.15[e] (0.27)
calc[f]	817	2715	4936	10679	2.035[g] (0.16)
exp[h]	354	2350	5400	10930	– (–)
exp[i]	–	–	4890	10850	– (–)
exp[j]	–	–	5000	10800	– (–)
exp[k]	–	–	–	–	1.975 (0.35)

Measured ΔE_Q are all at LNT

[a] Steffen et al. (1988), $T = 77$ K

[b] Hiroi and Takeda (1992)

[c] Taran and Langer (2001)

[d] Van Alboom et al. (1994)

[e] Experimental: Van Alboom et al. (1993) with 0.27 Fe apfu, $T = 80$ K

[f] Lin et al. (1993)

[g] Experimental: Lin et al. (1993) with 0.80 Fe apfu, T = 79 K; $\Delta E_Q = 2.047$ mm·s^{-1} at $T = 28$ K

[h] With 0.28 Fe apfu: Goldman and Rossman (1977)

[i] With 1.70 Fe apfu: Goldman and Rossman (1979)

[j] Average over five natural orthopyroxenes: Langer and Khomenko (1999)

[k] Wiedenmann et al. (1986), $T = 4.2$ K

The asymmetry parameter η(M2) = 0.23 is similar to that of the M1-site, η = 0.16. While the first value is in the same range as the experimental ones (Table 4.3), η(M1) is somewhat smaller though Mössbauer spectra are rather insensitive against variations of η with corresponding error margins about 0.1–0.2. In turn, calculated η values may be very sensitive against small geometrical variations, as has been demonstrated recently (Grodzicki et al. 2009). Hence, it cannot be excluded that for the low temperature geometry, not yet available,

η(M1) may be different. The calculated isomer shifts are δ(M1) = 1.28 mm·s^{-1} and δ (M2) = 1.25 mm·s^{-1} compared with the experimental values of 1.31 mm·s^{-1} and 1.26 mm·s^{-1} at 4.2 K (Regnard et al. 1986). Since isomer shifts for high-spin Fe^{2+} derived from SCC-Xα calculations are generally somewhat too small (Weber et al. 2009) the smaller difference of 0.03 mm·s^{-1} between the calculated values is assigned to δ(M2) being somewhat too large. As will be seen later when discussing the magnetic structure, this might be interpreted as an indication that the Fe(M2)–O3 distances are somewhat larger at RT than at temperatures below T_N. The sign of the EFG is positive for both iron sites. The EFG for Fe^{2+}(M1) is directed approximately along the crystallographic b axis, i.e., between the two oxygens O1A and O1B with the longest Fe–O distances of 2.194 Å (Fig. 4.2a). Accordingly, the occupied $3d\downarrow$ orbital is a linear combination of $3d_{xy}$ and $3d_{x^2-y^2}$ with respect to the principal axes system of the EFG at M1. The other four empty spin-down molecular orbitals with $3d$ character are mixtures of all five atomic $3d$-orbitals so that the assumption of a tetragonally compressed M1 octahedron with approximate symmetry D_{4h} (Steffen et al. 1988; Taran and Langer 2003) is not justified with regard to the electronic structure. The EFG for Fe^{2+}(M2) is oriented approximately along the c axis, i.e., parallel to the direction of the strong tetragonal compression (Fig. 4.2b). Accordingly, with regard to the principal axes system of the EFG at M2 the occupied $3d\downarrow$ orbital is an almost pure $3d_{x^2-y^2}$; and the four empty orbitals can be assigned in good approximation to atomic $3d$-orbitals, as well, with the ordering $3d_{xz}$, $3d_{yz}$, $3d_{xy}$ and $3d_z{}^2$ (Fig. 4.3). Decomposition of the EFG's into

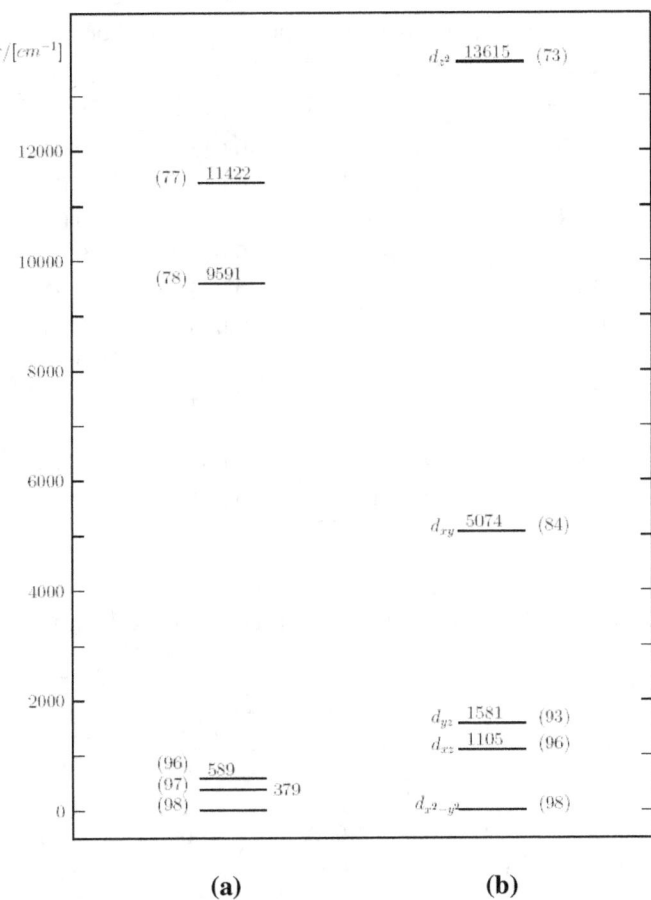

(a)   **(b)**

Figure 4.3 Orbital energy diagram of the $3d$ spin-down molecular orbitals for Fe^{2+}(M1) **(a)** and Fe^{2+}(M2) **(b)**. The numbers in brackets denote the Fe($3d$) percentage contribution to the respective molecular orbital. For Fe^{2+}(M2) the predominant atomic $3d$-character is given, as well, whereas

69

valence, covalence and ligand contributions shows that the difference in the quadrupole splittings arises mainly from the $4p$ valence and the covalence contributions that are -0.004 and -0.044 mm·s^{-1} for Fe^{2+}(M1) and -0.253 and -0.690 mm·s^{-1} for Fe^{2+}(M2), respectively, whereas the $3d$ valence contributions are very similar with $+3.205$ mm·s^{-1} (M1) and $+3.113$ mm·s^{-1} (M2), and the ligand contributions are almost negligible.

Table 4.4 Contributions to the overlap populations of the iron-oxygen bonds

	d (Å)	Total	$4s + 4p$	$3d\uparrow$	$3d\downarrow$
Fe(M1)–O$_{1A}$	2.083	0.203	0.185	-0.017	0.035
Fe(M1)–O$_{2A}$	2.088	0.229	0.207	-0.016	0.038
Fe(M1)–O$_{2B}$	2.123	0.194	0.178	-0.016	0.032
Fe(M1)–O$_{1B}$	2.124	0.181	0.166	-0.016	0.031
Fe(M1)–O$_{1A}$	2.194	0.144	0.134	-0.014	0.024
Fe(M1)–O$_{1B}$	2.194	0.149	0.139	-0.014	0.024
Fe(M2)–O$_{2B}$	1.993	0.301	0.264	-0.015	0.052
Fe(M2)–O$_{2A}$	2.023	0.254	0.220	-0.016	0.050
Fe(M2)–O$_{1B}$	2.128	0.167	0.145	-0.014	0.036
Fe(M2)–O$_{1A}$	2.157	0.153	0.133	-0.013	0.033
Fe(M2)–O$_{3A}$	2.459	0.037	0.037	-0.004	0.004
Fe(M2)–O$_{3B}$	2.596	0.041	0.041	-0.003	0.003

Table 4.5 Occupation numbers of molecular orbitals at various temperatures for size-converged 143 atom cluster around Fe^{2+}(M1) and 136 atom cluster around Fe^{2+}(M2) in orthoferrosilite

	M1			M2		
T (K)/ε_n(eV)	0.000	0.047	0.073	0.000	0.137	0.196
0	1.000	0.000	0.000	1.000	0.000	0.000
45	1.000	0.000	0.000	1.000	0.000	0.000
78	0.999	0.001	0.000	1.000	0.000	0.000
100	0.996	0.004	0.000	1.000	0.000	0.000
120	0.988	0.011	0.001	1.000	0.000	0.000
140	0.978	0.020	0.002	1.000	0.000	0.000
175	0.950	0.042	0.008	1.000	0.000	0.000
203	0.923	0.063	0.014	1.000	0.000	0.000
215	0.910	0.072	0.018	0.999	0.001	0.000
250	0.871	0.099	0.030	0.998	0.002	0.000
297	0.821	0.131	0.048	0.995	0.005	0.000
350	0.769	0.162	0.069	0.987	0.011	0.002
445	0.693	0.204	0.103	0.967	0.027	0.006

The calculated effective charges for Fe^{2+}(M1) of $+0.53$ and Fe^{2+}(M2) of $+0.64$, as well as those for the oxygen atoms between -0.35 and -0.43 exhibit substantial deviations from the formal oxidation states $+2$ and -2, respectively, and indicate considerable covalent character of the iron–oxygen bonds. A qualitative measure for the covalent part of a bond is the overlap population. The values listed in Table 4.4 show that about 90% of the total overlap populations arise from the orbital interactions of the (formally empty) iron 4s, 4p orbitals with the valence orbitals of the oxygens, whereas the completely filled $3d\uparrow$ orbitals yield a small negative, i.e., antibonding contribution because both the bonding and the antibonding linear combinations with the oxygen orbitals are occupied. The small contributions from the $3d\downarrow$ orbitals arise from the admixtures of the four formally empty $3d\downarrow$ orbitals to the bonding spin-down molecular orbitals, whereas the occupied $3d\uparrow$ -orbital has almost pure $3d$-character and does not contribute to the Fe-O bonds. Accordingly, the valence shell occupations of both Fe^{2+} contain considerable amounts from the 4s and 4p orbitals, viz. $Fe(M1)(4s^{0.52}4p^{0.55}3d^{6.41})$ and $Fe(M2)(4s^{0.46}4p^{0.46}3d^{6.43})$. The increased $3d$ occupations again arise from the admixtures of the four formally empty $3d\downarrow$ orbitals to the bonding spin-down molecular orbitals. Correspondingly, the $3d$ spin densities are slightly reduced from the free-ion value 4 to 3.55 and 3.51 for $Fe^{2+}(M1)$ and $Fe^{2+}(M2)$, respectively.

In the second series of model calculations, the temperature dependence of the quadrupole splitting and asymmetry parameter is simulated. It is assumed, that this dependence arises predominantly from thermal population of the low-lying unoccupied molecular orbitals and changes of the crystal structure are negligible (Geiger et al. 2003).

According to these assumptions, an evaluation of the thermal occupations of the excited orbitals at temperature T has exponential behavior $\exp\left(-\frac{\varepsilon_n - \varepsilon_o}{kT}\right)$ (Table 4.5). The occupation numbers for e_g-like orbitals in both clusters around $Fe^{2+}(M1)$ and $Fe^{2+}(M2)$ are below 10^{-4} for temperatures up to 445 K, due to the large d–d excitation energies. The resultant calculated values of the quadrupole splitting and asymmetry parameter are presented in Table 4.6.

Table 4.6 Calculated temperature dependence of the quadrupole splitting ΔE_Q and asymmetry parameter η for clusters with Fe^{2+} ion in the M1 and M2 sites

T, K	ΔE_Q, mm·s^{-1} (M1)	η (M1)	ΔE_Q, mm·s^{-1} (M2)	η (M2)
0	3.130	0.16	2.011	0.23
45	3.130	0.16	2.011	0.23
78	3.126	0.16	2.011	0.23
100	3.110	0.16	2.011	0.23
120	3.078	0.17	2.011	0.23
140	3.028	0.18	2.011	0.23
175	2.903	0.19	2.010	0.23
203	2.777	0.21	2.009	0.23
215	2.719	0.22	2.007	0.23
250	2.546	0.24	2.001	0.22
297	2.316	0.28	1.984	0.22
350	2.080	0.31	1.949	0.21
445	1.733	0.37	1.842	0.19

The small splitting between the highest occupied molecular orbital (HOMO) and the lowest unoccupied molecular orbital (LUMO) of 379 cm^{-1} for the cluster with central Fe^{2+}(M1) yields a pronounced temperature dependence of the quadrupole splitting and asymmetry parameter (Table 4.6). These dependences indicate a tendency that the direction of the EFG changes and deviates from axial symmetry.

In turn, the separation of 1105 cm^{-1} between the HOMO and the LUMO for Fe^{2+}(M2) is relatively large and results in a slight reduction of the quadrupole splitting. In addition, unlike to Fe^{2+}(M1), the asymmetry parameter is slightly decreasing with increasing temperature for Fe^{2+}(M2).

The calculated temperature dependence of the quadrupole splitting, which arises from thermal population of the low lying excited molecular orbitals, is in quantitative agreement with the experimental data presented in Table 4.7. The reliability of the proposed model in an earlier report is also confirmed by the obtained negative V_{ZZ}, whereas it is positive as the quadrupole splitting.

Table 4.7 Measured temperature dependence of the quadrupole splitting ΔE_Q for M1 and M2 sites

	$\Delta E_Q{}^a$, mm·s^{-1}			$\Delta E_Q{}^b$, mm·s^{-1}	
T, K	M1	M2	T, K	M1	M2
28	3.130	2.047	34	3.02	2.17
45	3.130	2.043	80	3.07	2.18
79	3.126	2.035	140	3.01	2.16
120	3.110	2.028	200	2.91	2.15
175	3.078	2.016	250	2.75	2.14
215	3.028	2.003	300	2.59	2.12
293	2.903	1.971	350	2.41	2.09
-	-	-	445	-	2.03

[a] Lin et al. (1993)

[b] Van Alboom et al. (1993)

Table 4.8 Calculated and measured d–d excitation energies (in cm^{-1}) and quadrupole splittings ΔE_Q (in mm·s^{-1}), with η in brackets, for a synthetic Mg-rich orthopyroxene (a) and a natural mantle orthopyroxene (b)

Atoms	$\varepsilon_1(\pi^*)$	$\varepsilon_2(\pi^*)$	$\varepsilon_3(\sigma^*)$	$\varepsilon_4(\sigma^*)$	$\Delta E_Q(\eta)$
143a (M1)	444	581	11962	14736	3.17 (0.23)
143b (M1)	460	920	12115	15325	2.97 (0.09)
136a (M2)	1033	1605	6033	14591	2.15 (0.20)
136b (M2)	645	1226	6332	14438	2.46 (0.20)

In the next series of model calculations, analogous calculations on a synthetic Mg-rich Opx, $(Mg_{0.75}Fe_{0.25})_2Si_2O_6$ (Yang and Ghose 1995), and a natural mantle Opx with small iron content (0.023 apfu in M1 and 0.127 apfu in M2) have been performed (Diego Gatta et al. 2007). Whereas the effective charges, occupation numbers of the valence shell of iron, $3d$ spin densities and overlap populations of the Fe–O bonds are similar for all Opx structures, the calculated spectroscopic data are considerably different (Table 4.8). While the calculated quadrupole splittings of 3.17 mm·s^{-1} ($\eta = 0.23$) and 2.97 mm·s^{-1} ($\eta = 0.09$) for Fe^{2+}(M1) in the synthetic Mg-rich Opx and the natural mantle Opx, respectively, are still close to the values in orthoferrosilite, the calculated quadrupole splittings of 2.15 mm·s^{-1} ($\eta = 0.20$) and 2.46 mm·s^{-1} ($\eta = 0.20$) for Fe^{2+}(M2) differ from the values in orthoferrosilite. Even more pronounced are the changes in the optical spectra due to the reduced average metal-oxygen distances in the species with low iron content. The splittings between the t_{2g}- and e_g-like orbitals increase to about 12000 and 15000 cm^{-1} at Fe^{2+}(M1) and to about 6200 and 14500 cm^{-1} at Fe^{2+}(M2) (Table 4.8).

These large changes are by far not reflected in the respective experimental data though a slight increase of the low energy d–d-transition at Fe^{2+}(M2) from 4910 to 5380 cm^{-1} is observed in going from orthoferrosilite to enstatite (Burns 1993). Altogether, these results confirm previous conclusions (Geiger et al. 2003; Weber et al. 2009) regarding the calculations of spectroscopic data of minerals with low iron content that are of limited reliability due to local distortions of the environment of iron that are not properly accounted for in experimental structure determinations by diffraction methods.

4.3 Magnetic Structure of Orthoferrosilite

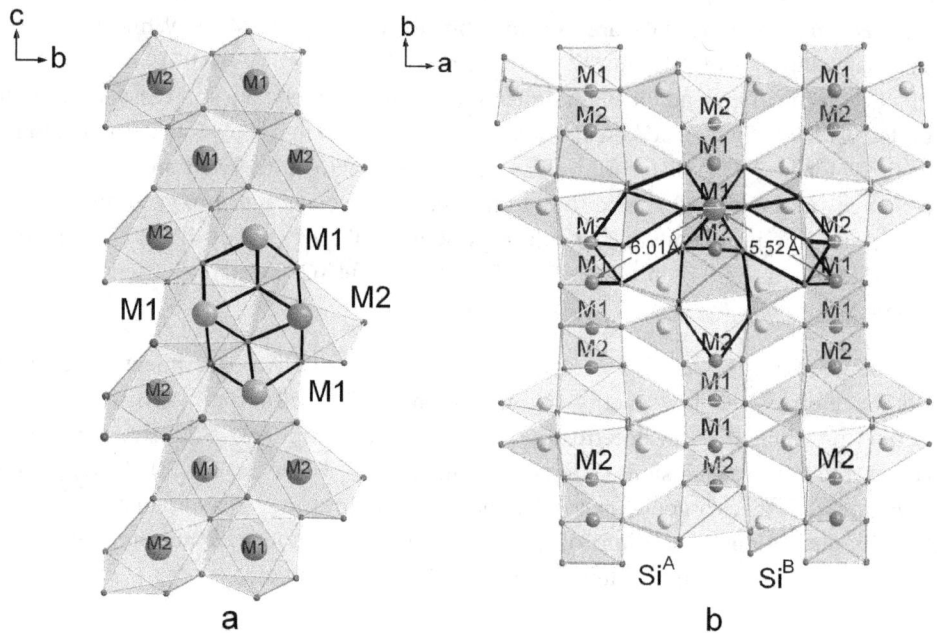

Figure 4.3 (a) - Magnetic exchange pathways (bold lines) between the iron spins in the (*b*, *c*)-plane within a ribbon. **(b)** -Magnetic exchange pathways (bold lines) between the iron spins of neighbouring ribbons in the (*a*, *b*)-plane.

Since orthoferrosilite consists of ribbons of edge-sharing M1 and M2 octahedra along the crystallographic *c* axis, two different intra-chain coupling constants, viz. J_{intra}(M1–M1) and J_{intra}(M1–M2), between two Fe^{2+} ions in neighbored octahedra (Fig. 4.3a) have to be calculated for a series of clusters of increasing size (Zherebetskyy et al. 2010). Since a low temperature structure is not yet available, all calculations had to be performed for the RT structure (Sueno et al. 1976), which may cause some limitations with regard to the reliability of the results, as will be discussed below. The smallest size-converged clusters contain 86 atoms with the composition Fe_2(M1)Fe_2(M2)O_{16}–$Si_{17}Mg_4O_7F_{38}$ for J_{intra}(M1–M1) and Fe_3(M1)Fe(M2)O_{16}–$Si_{16}Mg_6O_8F_{36}$ for J_{intra}(M1–M2), respectively. In these clusters, all divalent cations directly connected to one of the bridging oxygens are taken as Fe^{2+}. The calculated coupling constants collected in Table 4.9 indicate a fairly strong ferromagnetic coupling of almost equal size between both iron pairs within the ribbon. This result is in accordance with experiment (Wiedenmann et al. 1986), as well as with the expectation from the empirical Goodenough-Kanamori rules (Goodenough 1958, Kanamori 1959) since the bridging angles are close to 90°.

The pattern of the interchain interactions appears to be considerably more complicated (Fig. 4.3b). Decomposing first the total interaction between the ribbons into pair interactions, it is observed that the interactions of Fe(M2) with both Fe(M1) and Fe(M2) of neighboured ribbons are virtually zero, i.e., $|J| < 0.5$ cm^{-1}.

Table 4.9 Calculated spin-coupling constants J_{intra} between pairs of Fe^{2+}-ions within the same ribbon for clusters of 86 atoms

	M1–M1	M1–M2
d(Fe–Fe) [Å]	3.14	3.00
∠ (Fe, O, Fe)	95°, 93°	97°, 98°
J_{intra} (cm^{-1})	+16.0	+16.4

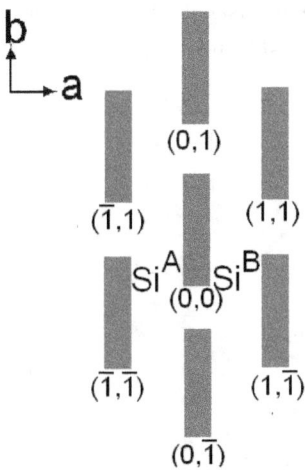

Figure 4.4 Schematic representation and nomenclature of neighbouring ribbons in the (a, b)-plane.

Since in both cases O_3 atoms are part of the various exchange pathways, this result has most likely to be attributed to the large Fe(M2)-O_3 distances. Consequently, the magnetic coupling between different ribbons has to be assigned to the interactions between the irons at the M1 site. As shown schematically in Fig. 4.4, the central ribbon at (0;0) may undergo three types of interactions, viz. (i) with a ribbon at ($\bar{1}$; $\bar{1}$) connected via Si^A tetrahedra with (0, 0); (ii) with a ribbon at (1; $\bar{1}$) connected via Si^B tetrahedra with (0;0); and (iii) with a ribbon at (0; $\bar{1}$) in the (b, c)-plane parallel to the ribbon at (0; 0). From the data displayed in Table 4.10, the following pattern of the spin structure emerges. At first, the coupling between the (0;0) and (1; $\bar{1}$)-ribbons via the Si^B tetrahedra, corresponding to the shortest Fe(M1)–Fe(M1) distance, is distinctly antiferromagnetic with $J_{inter} = -4.9$ cm^{-1}. The resulting ratio $|J_{inter}|/J_{intra} = 0.3$ is in quantitative agreement with experiment (Wiedenmann et al. 1986). Secondly, the interaction of the (0;0) -ribbon with the ribbon in the (b,c)-plane is weakly ferromagnetic with $J_{inter} = +2.2$ cm^{-1}, again in accordance with the experimental data derived from neutron diffraction experiments (Wiedenmann et al. 1986). However, the interaction between the ribbons connected by the Si^A-tetrahedra is ferromagnetic so that pairs of ferromagnetic (b,c)-planes are coupled antiferromagnetically via chains of Si^B-tetrahedra.

Although such a spin structure that may be denoted as "double-plane antiferromagnetic" (Fig. 4.5) resembles the pattern obtained recently for $LiFeGe_2O_6$ (Redhammer et al. 2009), it is at variance with the magnetic structure derived from the neutron diffraction data. According to these results each (b, c)-plane is coupled antiferromagnetically to both neighboured ones whereas the calculated spin structure corresponds to mode "L" in Wiedenmann et al. (1986) that has been excluded by these authors. Hence, the key problem of the calculations seems to be that they are based on the RT structure.

Table 4.10 Calculated spin coupling constants J_{inter} between neighboured ribbons short distance of 4.84 Å between Fe(M1) in $(0; 0)$ and Fe(M2) in $(\bar{1}; \bar{1})$

	$(0; 0)$- Si^A- $(\bar{1}; \bar{1})$	$(0; 0)$ - Si^B- $(1; \bar{1})$	$(0; 0)$ - Si^A, Si^B- $(0; \bar{1})$
d(Fe–Fe) [Å]	6.01	5.52	7.79
$\angle$ (Fe, O, O)	114°, 145°	127°, 135°	115°, 129°
			118°, 150°
N_{at}	168	190	138
J_{inter} (cm^{-1})	+3.4	–4.9	+2.2

The expected predominant geometrical change on cooling down should be a decrease of the two large Fe(M2)–O_3 distances (Smyth 1973; Sueno et al. 1976; Yang and Ghose 1995). As mentioned above, the vanishing coupling of the Fe(M2) spins with those on neighboured ribbons is attributed to these large distances, so that decreasing d(Fe(M2)–O_3) might lead to antiferromagnetic coupling between the $(0; 0)$- and $(\bar{1}; \bar{1})$-ribbons, as well, especially in view of the short distance of 4.84 Å between Fe(M1) in $(0; 0)$ and Fe(M2) in $(\bar{1}; \bar{1})$.

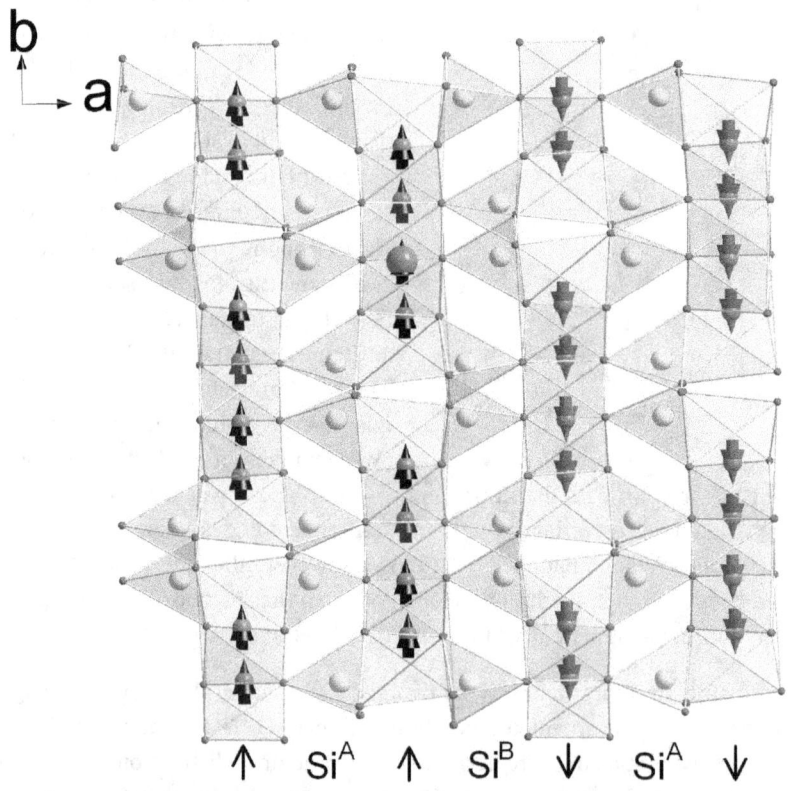

Figure 4.5 Calculated spin pattern within the (a, b)-plane.

4.4 Summary

The electronic and magnetic structure of the chain silicate orthoferrosilite has been characterized by electronic structure calculations in the LSDA (Zherebetskyy et al. 2010). The calculated spectroscopic data are in quantitive agreement with the corresponding experimental values, allowing the conclusion that the electronic structure of orthoferrosilite is correctly described, at least qualitatively. The magnetic interactions between the iron spins within the ribbons and between neighboring ribbons in the (b, c)-plane are ferromagnetic in accordance with the experimental results. The coupling between different (b, c)-planes is antiferromagnetic via Si^B tetrahedra but ferromagnetic via the Si^A tetrahedra. This ferromagnetic coupling, however, is at variance with an experimentally determined magnetic structure. It is argued that this discrepancy may be attributed to differences in the Fe(M2)–O_3 distances at room temperature, that had to be used in the calculations, and the geometry below the Néel temperature. Hence, as soon as low temperature structural data are available, this problem may be resolved.

Chapter 5

ELECTRONIC AND MAGNETIC STRUCTURE OF ALMANDINE

5.1 Introduction

Almandine, $Fe^{2+}_3Al_2Si_3O_{12}$, is the ferrous iron end member of the class of garnet minerals representing an important group of rock-forming silicates, which are the main constituents of the Earth´s crust, upper mantle and transition zone; therefore, a detailed understanding of its physical properties is essential. Almandine garnet crystallizes as other garnets in the cubic space group $Ia\bar{3}d$ (Figure 5.1), with unit-cell parameter $a = 11.512$ Å at 100 K (Geiger et al. 1992). The unit cell of almandine consisting of 160 atoms has fractional coordinates presented in Table 5.1 and the respective coordinates of equivalent positions for the space group $Ia\bar{3}d$, No. 230, are presented in Chapter 6.

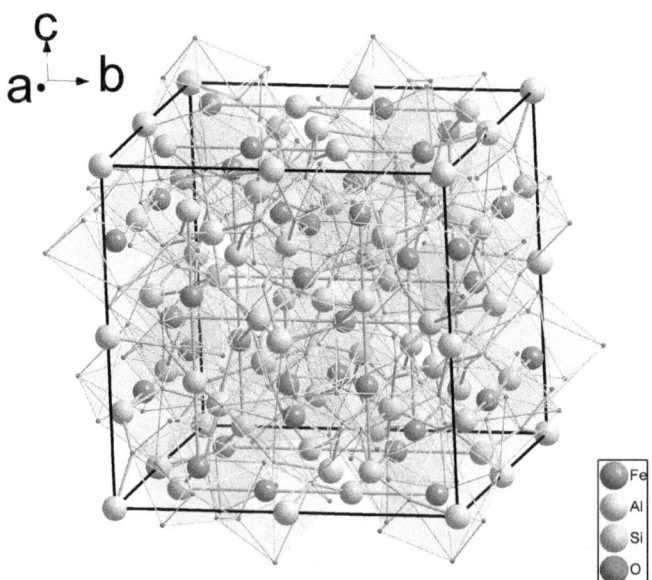

Figure 5.1 Unit-cell of almandine.

Table 5.1. Fractional atomic coordinates for synthetic almandine at 100 K (Geiger et al. 1992)

Atom	Oxidation	x	y	z
Fe	+2	0.00000	0.25000	0.12500
Al	+3	0.00000	0.00000	0.00000
Si	+4	0.37500	0.00000	0.25000
O	−2	0.03395	0.04943	0.65268

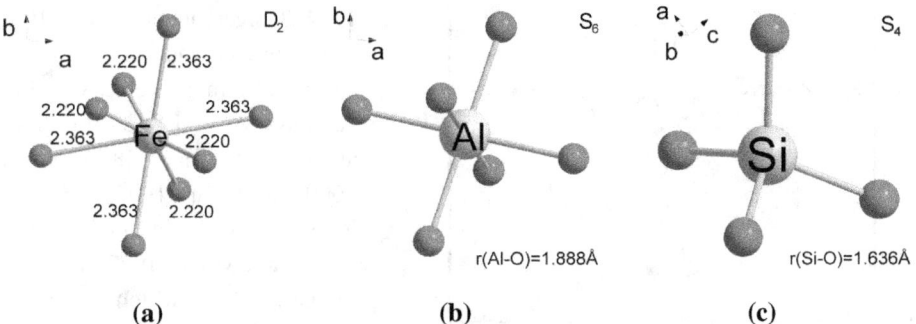

(a) (b) (c)

Figure 5.2 Cation polyhedra and their symmetry in the structure of Almandine: **(a)** - Fe dodecahedron, **(b)** - Al octahedron, **(c)** - Si tetrahedron.

Within the structure of almandine, the Fe^{2+} cations are coordinated to eight oxygens in a triangular dodecahedral arrangement, the Al ions occupy sixfold coordinated octahedral sites and Si occurs in tetrahedral coordination with four oxygens (Fig. 5.2) (Novak and Gibbs 1971, Geiger et al. 1992). Every oxygen atom is coordinated by one Al, one Si and two Fe^{2+} cations. In this structure, the direct environment of the Fe-dodecahedron consists of two edge-shared and four corner-shared Si tetrahedra; four edge-shared Al-octahedra and four edges of the Fe-dodecahedron are shared with other four dodecahedra with the next-nearest-neighbour Fe–Fe distance of 3.525 Å (Fig. 5.3). Each pair of Fe^{2+} ions is connected by two identical paths via oxygen bridges forming Fe–O–Fe angles of 101°.

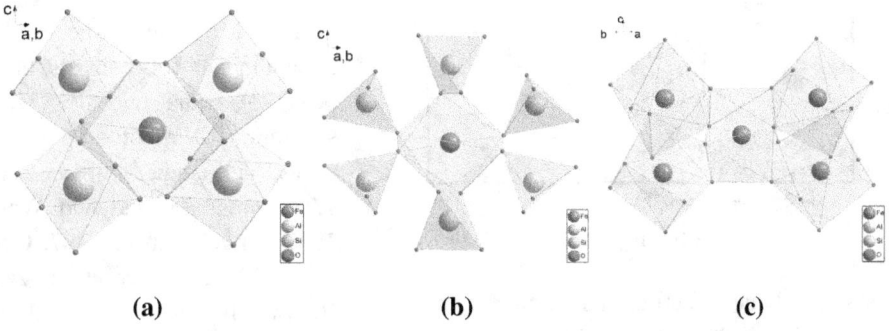

(a) (b) (c)

Figure 5.3 Neighbouring: **(a)** – Al, **(b)** – Si, **(c)** – Fe cations and polyhedra of Fe cation.

Almandine and other members of the $Fe^{2+}_3Al_2Si_3O_{12}$-$Mg_3Al_2Si_3O_{12}$ join have already been studied experimentally by several workers by absorption spectroscopy (Clark 1957, Geiger and Rossman 1994, Manning 1967, White and Moore 1972, Taran and Langer, 2001), Mössbauer spectroscopy and measurements of the magnetic properties (Murad and Wagner 1987, de Oliveira et al. 1987 and 1989, Geiger et al. 1992, Anovitz et al. 1993, Woodland and Ross 1994, Geiger and Feenstra 1997). The optical and near-to-mid-IR spectra of almandine show a three-band system in the near-infrared at about 4350 cm^{-1}, 5850cm^{-1} and 7600 cm^{-1} (Clark 1957, Geiger and Rossman 1994, Manning 1967, White and Moore 1972, Taran and Langer, 2001) which have been attributed to spin-allowed d–d transitions of Fe^{2+} from the spin-down d_z^2 orbital occupied in the ground state to the d_{yz}, $d_{x^2-y^2}$ and d_{xz} orbitals (Geiger et al. 2003). From the temperature dependence of the quadrupole splitting ΔE_Q as measured by Mössbauer spectroscopy a splitting pattern of the spin-down e_g-like orbitals has been derived as 1100 cm^{-1} (d_{xy}) with respect to the occupied d_z^2-orbital (Huggins 1975, Geiger et al. 2003).

The Mössbauer spectrum of almandine at RT consists of two peaks of equal intensity, with a separation ΔE_Q of about –3.50 $mm \cdot s^{-1}$ and isomer shift δ of about 1.28 $mm \cdot s^{-1}$ (Murad and Wagner 1987, Oliveira et al. 1987, Geiger at al. 1992, Geiger et al. 2003). Below 10 K the quadrupole doublet broadens within a temperature of few degrees. Rapidly, at 9.5 K, a widely split

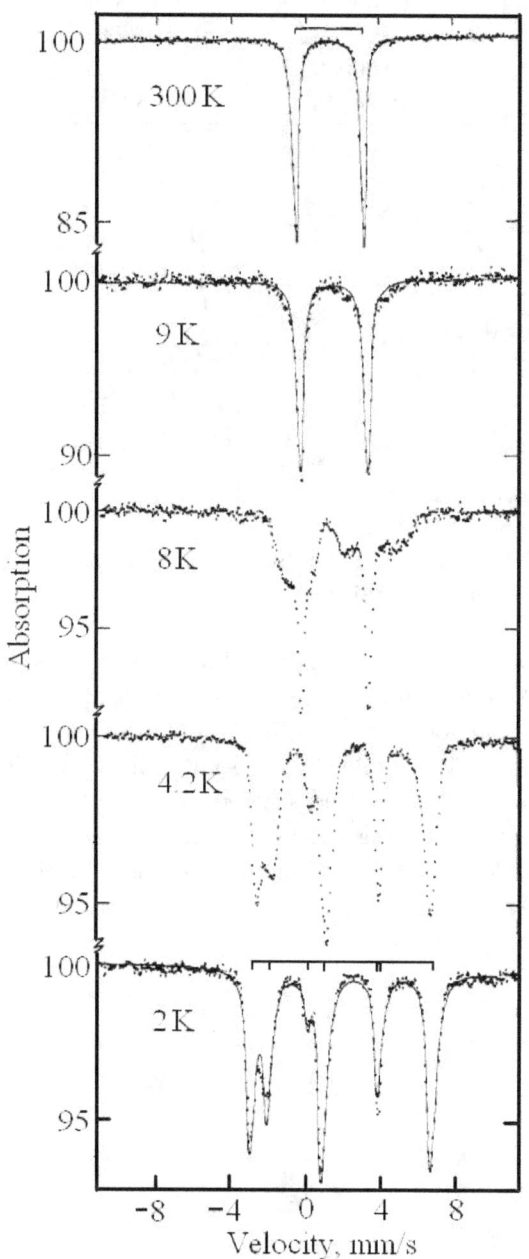

Figure 5.4 Mössbauer spectra of almandine at various temperatures. The shifts are shown by the solid lines (Oliveira et al. 1987).

pattern develops which means that almandine orders magnetically at low temperatures (Murad and Wagner 1987). At 2 K almandine exhibit a complex spectrum of two eight-line magnetic hyperfine patterns (Fig. 5.4) of equal intensities with nearly identical quadrupole splittings that average -3.70 mm·s^{-1} (de Oliveira 1987, 1989, Murad and Wagner 1987). Possible interpretations of the two magnetic hyperfine patterns, as assumed so far, are the existence of two different polar angles between the z-axis of the EFG and the direction of the magnetic hyperfine field, or the existence of two distinctly different EFG's (Murad and Wagner 1987, de Oliveira et al. 1989), or deviations from cubic symmetry common in garnets (Murad and Wagner 1987). Magnetic susceptibility (Prandl 1971) and low-temperature heat capacity (Anovitz et al. 1993) measurements reveal that the ordering at low temperatures is antiferromagnetic with $T_N = 7.5$ K. Magnetization measurements also provide an effective magneton number of 4.71 μ_B (Oliveira et al. 1987) for Fe^{2+} ions in almandine structure.

In the present research, the iron end member almandine is investigated, in order to arrive at a more detailed understanding of the electronic and magnetic structure of garnets. The calculations are based on clusters of increasing size that model the local environment of the Fe^{2+} sites. The reliability and suitability of this theoretical approach are assessed by comparing the computed spectroscopic data with the experimental results. Subsequently, the magnetic structure is derived by calculating the Heisenberg coupling constants for the various possible interactions between Fe^{2+} cations within this mineral.

The construction of suitable model clusters based on the experimentally determined geometry of a synthetic single crystal of almandine at 100 K (Geiger et al. 1992) follows the experience from previous cluster calculations on minerals (Grodzicki et al. 2009 and references therein).

5.2 Electronic Structure of Almandine

In general, size-converged results for optical transitions, hyperfine parameters and magnetic coupling constants are obtained if all coordination polyhedra of the cations bonded to the oxygen atoms of the first coordination sphere of the central transition metal ion are included though occasionally smaller clusters may suffice.

In the case of the iron ion of almandine, the cations of the second coordination sphere comprise six (formally tetravalent) Si ions, four (formally trivalent) Al ions and four divalent Fe cations. These ions are fourfold (Si), sixfold (Al) and eightfold (Fe) coordinated, respectively, by oxygen atoms. The inclusion of these polyhedra yields the cluster [Fe^{2+}O$_8$-Si$_6$Al$_4$Fe$_4$O$_{40}$]$^{50-}$ with 40 oxygen atoms in the third shell. Terminating the cluster at these oxygens, however, supplies an inappropriate description of the electronic properties. As it was mentioned in Section 2.5, first of all, a negative cluster charge has to be compensated by a Madelung-type potential, e.g., of appropriately distributed point charges. Secondly, some or all of the incompletely coordinated terminal oxygens have to be replaced with fluorine atoms, or completely omitted, if the metal-oxygen distances are too large. Such a procedure shifts the 2p orbitals into the energy range typical for saturated O(2p) orbitals and leads to a significant reduction of the cluster charge. In addition, Fe ions in more distant polyhedra are substituted by Mg ions because the resulting changes in the computed spectroscopic data are within the error margins expected due to the general theoretical approximations. Hence, the resulting cluster, that contains all coordination polyhedra of the cations bonded to the oxygen atoms of the first coordination sphere of the central iron ion (Fig. 5.5) constructed along these rules, comprises 55 atoms and has the composition $[Fe^{2+}O_8\text{-}Si_6Al_4Mg_4F_{32}]^{2-}$. The valence basis set consists of 2s, 2p orbitals for O and F, 3s, 3p orbitals for Mg, Si and Al, and 3d, 4s, 4p orbitals for Fe.

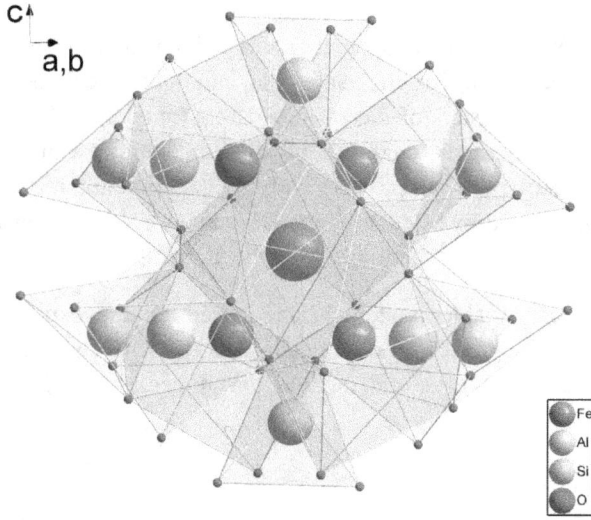

Figure 5.5 The 1st, 2nd and 3rd coordination spheres of the central Fe^{2+} ion.

The subsequent stepwise extension of the cluster size is realized by adding of the corresponding neighbouring polyhedra and the simultaneous replacing of oxygen and iron atoms by fluorine and magnesium atoms, respectively, in such a way that the cluster charge always remains –2 and the local symmetry of the central Fe^{2+} ion is preserved.

The largest size-converged cluster used for the calculation of the spectroscopic data of iron contains 189 atoms (Fig. 5.6) with the composition $[Fe^{2+}O_8\text{-}(Si_6Al_4Mg_4O_{40})\text{-}(Si_{16}Al_{12}Mg_{22}O_{20}F_{56})]^{2-}$. This cluster comprises all atoms around the central iron within a sphere of 7.68 Å, i.e. all cations are included up to this distance, and within 6.37 Å all anions are correctly represented as oxygen atoms. This cluster contains 761 valence orbitals and 998 valence electrons, and thus is a large system in a quantum chemical sense.

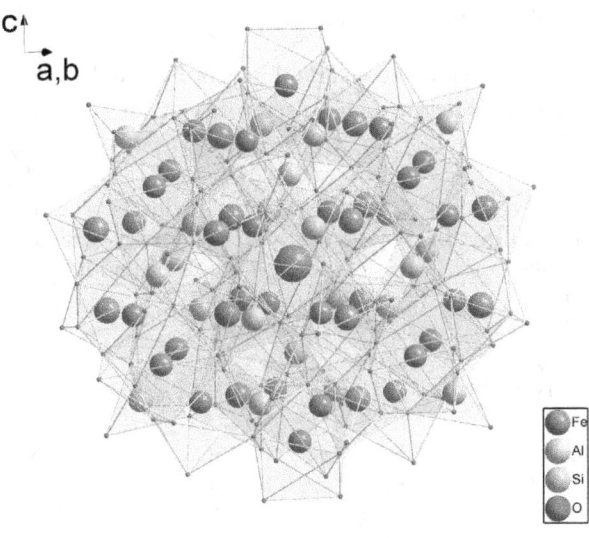

Figure 5.6 The largest cluster describing electronic structure of almandine contains 189 atoms.

If the polyhedron around Fe^{2+} were regular with Fe-O distances of 2.292 Å, the five spin-down $3d$ orbitals of Fe^{2+} are split by a cubic ligand-field into a lower doublet of e_g-orbitals and an upper triplet of t_{2g}-orbitals with the separation energy of 2767 cm^{-1} in accordance with the CFT (Section 3.1). Actually, the first coordination sphere of Fe^{2+} in almandine is regarded as the triangular dodecahedron with D_2 symmetry (Fig. 5.2 a). The splitting pattern of the five spin-down $3d$ orbitals of Fe^{2+} as obtained for the correspondent cluster $[Fe^{2+}O_8]^{14-}$ contains the e_g–like orbitals (ε_0 and ε_1), which are split by 670 cm^{-1} where the ε_0 orbital is the lowest and only occupied orbital, and the t_{2g}–like orbitals (ε_2, ε_3 and ε_4), which are split by almost 700 cm^{-1} and 1800 cm^{-1} (corresponding to 1887, 2589 and 3694 cm^{-1} with respect to the ε_0 orbital). These splitting of e_g and t_{2g} orbitals reflects the difference between the Fe-O distances and deviation of O-Fe-O angles from 90° within the dodecahedron. The calculated quadrupole splitting is −3.18 $mm·s^{-1}$ (η=0.16).

Adding all polyhedra of the second coordination sphere increases the splitting of the e_g –like orbitals to 903 cm^{-1}. The splitting of the t_{2g}–like orbitals shows the same behaviour with energy differences between ε_3 and ε_4 orbitals of 1444 cm^{-1} and 4033 cm^{-1} with respect to the ε_2 orbital. For the resulting 55 atom-containing cluster, the calculated quadrupole splitting increases to −3.61 $mm·s^{-1}$. The asymmetry parameter increases from η=0.16 to η=0.23. This is the smallest reasonable cluster providing reasonable results for spectroscopic data.

Next, expanding the cluster from 55 up to 111 atoms farther increases the splitting between e_g–like orbitals, and between e_g and t_{2g} orbitals, but does not lead to significant changes of the general picture. Subsequent increasing of the cluster size from 111 up to 189 atoms does not change appreciably the d-d-excitation energies or quadrupole splitting and asymmetry parameter (Table 5.2), so that a cluster of 111 atoms appears to be size convergent with respect to the calculated spectroscopic data.

The splitting scheme of the spin-down $3d$ orbitals of Fe^{2+} in the crystal fields of the regular cube, in the triangular dodecahedron with D_2 symmetry and in the almandine crystal structure is presented in Fig. 5.7.

Table 5.2 Calculated d-d excitation energies (in cm^{-1}) with respect to the lowest $3d$-orbital (ε_0) and quadrupole splitting ΔE_Q (in $mm \cdot s^{-1}$ with asymmetry parameter η in brackets) for almandine clusters of increasing size around Fe^{2+} ion

Atoms	$\varepsilon_1(\sigma^*)$	$\varepsilon_2(\pi^*)$	$\varepsilon_3(\pi^*)$	$\varepsilon_4(\pi^*)$	$\Delta E_Q(\eta)$
9	670	1887	2589	3694	-3.18 (0.16)
31	339	1799	2630	4735	-3.86 (0.15)
55	903	2742	4186	6775	-3.61 (0.23)
67	1153	3057	4154	5485	-3.23 (0.20)
83	1178	3170	4364	5832	-3.37 (0.18)
111	1387	3049	4864	7364	-3.41 (0.22)
127	1395	3114	4936	7477	-3.45 (0.23)
129	1323	3291	4969	7340	-3.50 (0.18)
161	1355	3404	4920	6985	-3.39 (0.19)
165	1307	3122	4638	6630	-3.48 (0.21)
177	1363	3372	4968	7042	-3.54 (0.20)
189	1331	3581	5065	7009	-3.42 (0.16)

In the next step, in order to assess the reliability of the theoretical approach and the suitability of the model clusters, the calculated spin-allowed d-d transitions and quadrupole splitting for the 189-atom cluster around Fe^{2+}-cation are compared with the respective experimental data

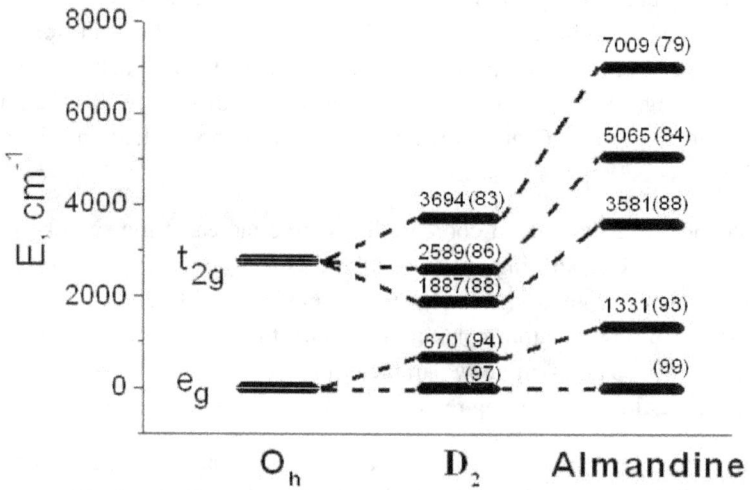

Figure 5.7 Spin-down $Fe^{2+}(3d)$ orbital energies (in cm^{-1}) relative to the lowest occupied orbital for O_h, D_2 and Almandine crystal fields together with the Fe(3d) percentage contribution in parentheses.

(Table 5.3) obtained by optical absorption and Mössbauer spectroscopy (Clark 1957, Geiger and Rossman 1994, White and Moore 1972; Murad and Wagner 1987, Geiger et.al. 1992, Woodland and Ross 1994, Geiger and Feenstra 1997).

Table 5.3. Calculated and measured d-d excitation energies (in cm^{-1}) with respect to the lowest $3d$-orbital (ε_0), quadrupole splittings ΔE_Q (in $mm \cdot s^{-1}$) with η in brackets and isomer shifts δ (in $mm \cdot s^{-1}$) for a size converged 189 atom cluster around Fe^{2+} ion in almandine

Nat	$\varepsilon_1(\sigma^*)$	$\varepsilon_2(\pi^*)$	$\varepsilon_3(\pi^*)$	$\varepsilon_4(\pi^*)$	$\Delta E_Q(\eta)$	δ
theo[o]	1331	3581	5065	7009	-3.42 (0.16)	1.28
exp[a]	--	2900	6000	7900	--	
exp[b]	1100	4317	5733	7564	--	
exp[c]	<2000	4381	5874	7663	--	
exp[d]	--	4500	6000	7800	--	
exp[e]	--	4340	5850	7650	--	
exp[f]	--	--	--	--	-3.66	1.28
exp[g]	--	--	--	--	-3.65	1.28
exp[h]	--	--	--	--	-3.54	1.31
exp[i]	--	--	--	--	-3.67	1.27
calc[j]	1687	4363	6299	7928	-3.67 (0.03)	1.27

[o] This work
[a] Clark (1957)
[b] Geiger and Rossman (1994)
[c] White and Moore (1972)
[d] Manning (1967)
[e] Taran and Langer (2001)

[f] Murad and Wagner (1987)
[g] Geiger et al. (1992)
[h] Woodland and Ross (1994)
[i] Geiger and Feenstra (1997)
[j] Geiger et al. (2003)

The comparison reveals that the results are in quantitative agreement for the quadrupole splitting and isomer shifts obtained by Mössbauer spectroscopy (Murad and Wagner 1987, Geiger et al. 1992, Woodland and Ross 1994, Geiger and Feenstra 1997) within the expected error margins of the calculations, whereas the d-d energies are somewhat smaller than the respective experimental values (Clark, 1957, Geiger and Rossman, 1994, White and Moore 1972, Manning, 1967).

In addition, the calculated magnetic moment of 4.40 μ_B for Fe^{2+} ions in almandine is in quantitative agreement with the experimental value of 4.71 μ_B (Oliveira et al. 1989) within the expected error margins of the calculations, where μ_B is Bohr magneton. The value of the magnetic moment of Fe^{2+} ion in almandine is smaller than the one for the ideal Fe^{2+} ion of 4.9

μ_B due to the reduction of the spin density of Fe^{2+} cation caused by covalent interaction with surrounding anions.

The calculations of effective charges are useful to understand the chemical bond between iron and neighbouring oxygen atoms in almandine. The calculated effective charges for iron atoms of +0.70 and for the oxygen atoms are between –0.44 and –0.46 showing substantial deviations from the formal oxidation states of +2 and –2, respectively. Such deviations indicate the considerable covalent character of the Fe–O chemical bonds within almandine structure. The occupation of the valence shells of iron $4s^{0.52}4p^{0.31}3d^{6.47}$ reveals that this covalent character is mainly due to the 4s- and 4p- electrons whereas the *3d*-shell occupation number is moderately increased and therefore the contribution of the *3d*- electrons to the Fe-O bonds is relatively small.

5.3 Magnetic Structure of Almandine

Since in almandine every Fe dodecahedron is edge-shared with four identical Fe dodecahedra, the coupling constants between Fe^{2+} ions in these neighboured dodecahedra (Fig. 5.3 c) have to be calculated first for a series of clusters of increasing size. Since crystal structure data for the temperatures below Néel temperature are not available, all calculations are performed for the 100 K structure (Geiger et al. 1992).

Generally, the starting point for the construction of a cluster for interacting iron ions is the smallest cluster providing reasonable results for spectroscopic data. As mentioned in Section 5.2, the smallest cluster must include all coordination polyhedra of the cations bonded to the oxygen atoms of the first coordination sphere of the central transition metal ion. In this case, the 55 atom-containing cluster (Fig. 5.5) satisfies these requirements and is the smallest reasonable cluster (Table 5.2).

As noticed in Section 3.4, in order to derive and explain the spin structure of almandine, the possible exchange pathways have to be discussed and analysed. Using these exchange pathways, respective clusters have to be constructed and investigated.

Firstly, it should be emphasized, that every pair of the edge-shared Fe dodecahedra are identical in almandine. Secondly, there are two possible interaction paths via two bridging oxygens between each pair of Fe cations in the edge-shared dodecahedra (Fig. 5.8a). These two pathways form Fe–O–Fe angle of 101^0 with Fe–O bonding lengths of 2.220 and 2.363 Å. Next, every pair of the edge-shared dodecahedra has connections to third common dodecahedron (Fig. 5.8b). Finally, as mentioned in Section 5.1, every dodecahedron has four common edges with other four dodecahedra. On this basis, three clusters describing the superexchange interaction between adjacent Fe^{2+} ions are constructed (Fig. 5.8). These clusters comprise series of the magnetic structure calculations which investigate the cluster size dependence of the magnetic coupling between two, three and five adjacent Fe^{2+} ions of edge-shared dodecahedra.

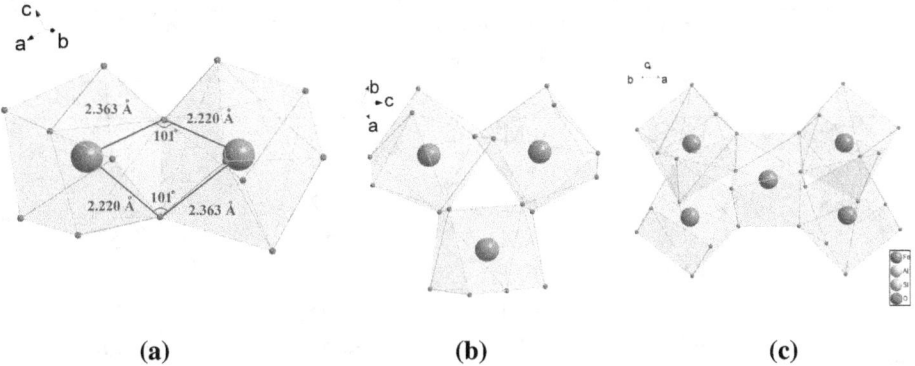

(a)   **(b)**   **(c)**

Figure 5.8 The adjacent interacting Fe^{2+} ions in the magnetic structure

From the MO-calculations, a weak ferromagnetic coupling with $J = +3.52$ cm^{-1} is obtained between the spins of the two nearest iron ions (Fig. 5.8 a) via the bridging oxygens for the 79 atom cluster with the composition Fe_2O_{14}-$Mg_5Al_6Si_{10}O_2F_{40}$, comprising of 326 valence orbitals and 460 valence electrons, which is based on the smallest reasonable cluster (Section 5.2). The quadrupole splitting values for the two iron atoms in ferromagnetic (FM) and antiferromagnetic (AM) configurations are equal (Table 5.4) reflecting their identical atomic environment and equal lattice sites. Furthermore, the value of the quadrupole splitting of -3.53 mm·s^{-1} is in quantitative agreement with the Mössbauer measurements (Murad and Wagner 1987, Geiger et.al. 1992, Woodland and Ross 1994, Geiger and Feenstra 1997).

Subsequent increasing of the cluster size from 79 to 118 atoms leads to oscillatory changes of the values of the Heisenberg coupling constant and quadrupole splitting. Nevertheless, the quantities of quadrupole splitting slightly improve in accordance with the experimental data, while the spin coupling remains positive and weak. Further increasing of the cluster size from 118 to 136 atoms does not change the common picture appreciably (Table 5.4), so that the cluster of 118 atoms appears to be size convergent with respect to the calculated data.

In the second series of magnetic structure calculations, describing the magnetic coupling between three adjacent Fe^{2+} ions of edge-shared dodecahedra (Fig. 5.8 b), the smallest neutral cluster contains 89 atoms with the composition $(Fe^{2+}O_6)_3$-$Si_{12}Al_8Mg_6O_{12}F_{30}$ and comprises 371 valence orbitals and 498 valence electrons. The subsequent stepwise extension of the sizes of the clusters is realized in such a way that the charges of the clusters always remain neutral.

The Heisenberg coupling constant increases up to +5.87 cm^{-1} for the cluster, while the values of the quadrupole splitting change to -3.32 mm·s^{-1}. However, they are still in satisfactory agreement with the Mössbauer measurements within the expected error margins of the calculations.

Table 5.4 Calculated spin-coupling constants and quadrupole splittings between two iron ions for almandine clusters of increasing size around Fe^{2+} ions

N_{at}	J, cm^{-1}	Magnetic configuration	ΔE_Q, mm·s^{-1}	
			1	2
79	+3.52	FM	-3.53	-3.53
		AM	-3.53	-3.53
88	+5.1	FM	-3.52	-3.52
		AM	-3.53	-3.53
102	+2.31	FM	-3.63	-3.63
		AM	-3.63	-3.63
112	+4.1	FM	-3.52	-3.52
		AM	-3.53	-3.53
118	+1.87	FM	-3.58	-3.58
		AM	-3.59	-3.59
128	+1.46	FM	-3.61	-3.61
		AM	-3.61	-3.61
136	+2.1	FM	-3.6	-3.61
		AM	-3.62	-3.62

Expanding the cluster size up to 113 atoms leads to slight improvement of the quadrupole splitting in accordance with the experimental data and the coupling constant decreases to +4.95 cm^{-1}. Further extension of the cluster size up to 137 atoms does not change the quadrupole splitting or the Heisenberg coupling constant, so that the cluster of 113 atoms is size convergent (Table 5.5).

Table 5.5 Calculated spin-coupling constants and quadrupole splittings between three iron ions for almandine clusters of increasing size around Fe^{2+} ions

N_{at}	J, cm^{-1}	Magnetic configuration	ΔE_Q, mm·s^{-1}		
			1	2	3
89	+5.87	FM	-3.32	-3.32	-3.32
		AM	-3.31	-3.31	-3.32
113	+4.95	FM	-3.38	-3.38	-3.38
		AM	-3.36	-3.36	-3.37
137	+4.96	FM	-3.39	-3.39	-3.39
		AM	-3.37	-3.37	-3.39

Finally, the interaction between the five nearest iron atoms is also ferromagnetic with a spin coupling constant of +3.56 cm^{-1} and satisfactory values of the quadrupole splitting with regard to the experimental values (Table 5.6). The quantity of quadrupole splitting for the central iron atom is not equal to other irons because of different atomic environments: the central iron is surrounded by the four nearest iron atoms, whereas these four irons have in their environment only the two nearest iron atoms (Fig. 5.8 c).

Table 5.6 Calculated spin-coupling constant and quadrupole splittings between

five iron ions

N_{at}	J, cm^{-1}	Magnetic configuration	ΔE_Q, mm·s^{-1}				
			1	2	3	4	5
143	+3.56	FM	-3.44	-3.52	-3.52	-3.52	-3.52
		AM	-3.60	-3.56	-3.56	-3.56	-3.56

Accordingly, the positive values of the spin coupling constants (Tables 5.4-5.6) provide information on the ferromagnetic mode of the interaction between the spins of nearest iron ions in edge-shared dodecahedra. These results are in accordance with the expectation from the empirical Goodenough-Kanamori rules (Goodenough 1958, Kanamori 1959), since the bridging Fe–O–Fe angles of 101^0 are close to $90°$.

However, experimental measurements reveal that the ordering at low temperatures is antiferromagnetic. Hence, the following question arises naturally: where is antiferromagnetic interaction within the almandine structure? The answer to this question requires a careful consideration of almandine crystal structure.

The Fe edge-shared dodecahedra form a 3D lattice (Fig. 5.9) of circles of 10 edge-shared dodecahedra (Fig. 5.10). As shown above, the spins of iron ions of edge-shared dodecahedra interact ferromagnetically; therefore, the spins of the iron ions within this 3D lattice are ferromagnetically coupled.

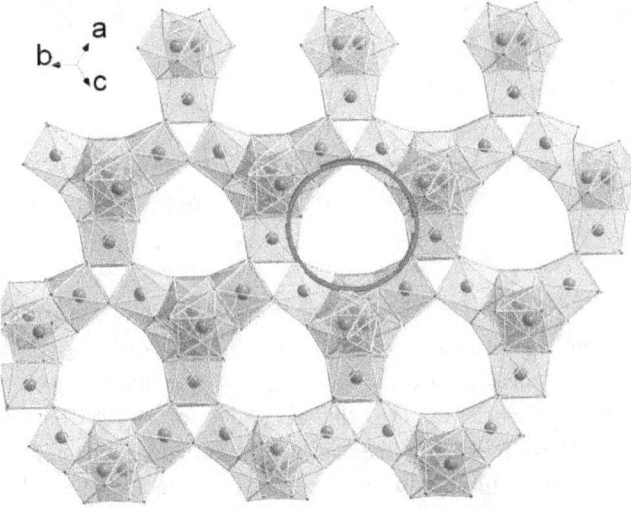

Figure 5.9 (111) projection of the 3D lattice formed by the Fe edge-shared dodecahedra.

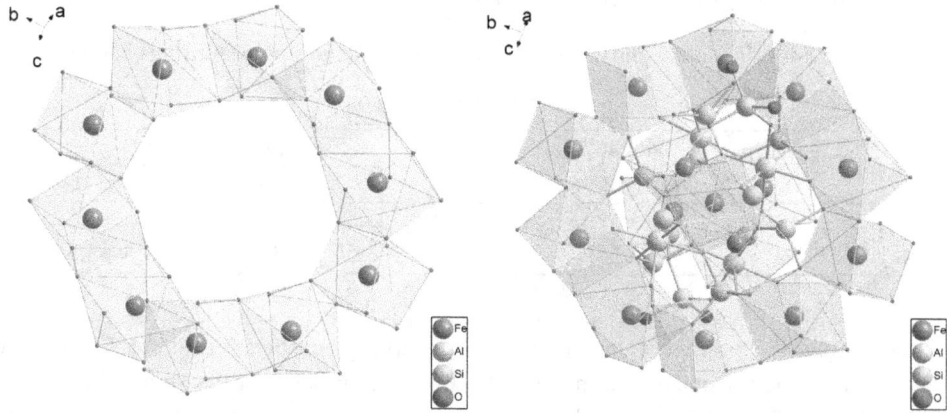

Figure 5.10 Circle of the 10 Fe edge-shared dodecahedra.

Figure 5.11 Circle with its interior.

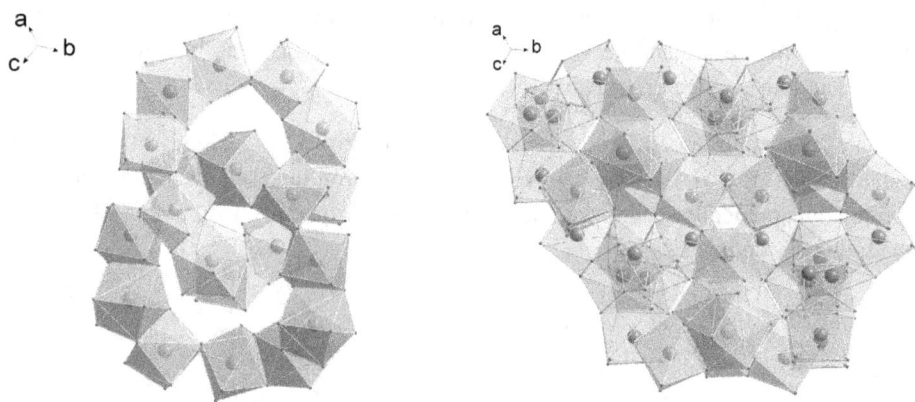

Figure 5.12 Two interpenetrating circles.

Figure 5.13 (111) projection of two interpenetrating crystal sublattices.

Further consideration of the circle's interior reveals that it contains Al octahedra and Si tetrahedra, which in turn, are connected to other Fe dodecahedra (Fig. 5.11). These inner Fe dodecahedra are a part of another circle (Fig. 5.12), which is connected to the first circle via Si tetrahedra and Al octahedra. In turn, this second circle is the part of another second sublattice (Fig. 5.13). These two sublattices are connected via Si tetrahedra and Al octahedra. Hence, the magnetic interaction between iron ions connected via Al octahedra and Si tetrahedra have to be calculated. For this purpose, two clusters, describing the interaction between the 6 Fe ions via Si tetrahedra (Fig. 5.14 a) and Al octahedra (Fig. 5.14 b), are constructed. Both clusters contain 165 atoms with the composition $Fe^{2+}_6O_{40}$-$Si_{23}Al_{12}Mg_{12}O_{12}F_{60}$ for the interaction via Si tetrahedra and $[Fe^{2+}_6O_{36}$-$Si_{18}Al_{15}Mg_{12}O_6F_{72}]^{-3}$

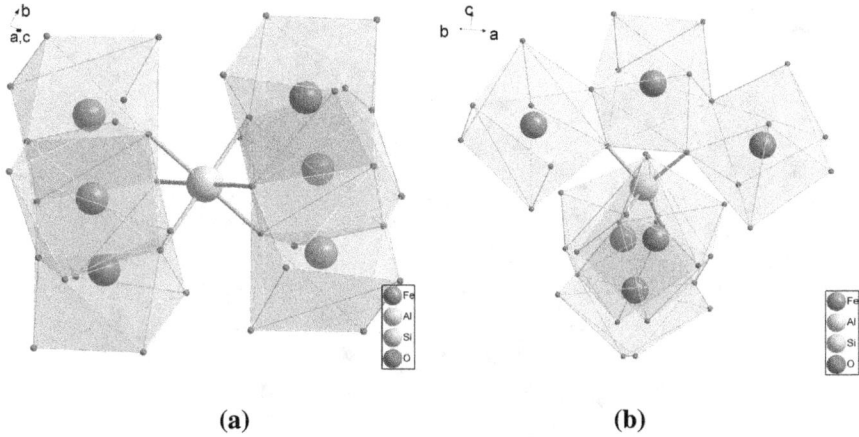

Figure 5.14 The 6 Fe dodecahedra connected via:
(a) – Si tetrahedron, **(b)** – Al octahedron.

for the interaction via Al octahedra, respectively. The couplings between the spins of iron ions via both Si tetrahedra and Al octahedra are weakly antiferromagnetic with $J = -0.07$ cm^{-1} and $J = -0.78$ cm^{-1} (Table 5.7). The smaller value of the spin coupling constant for magnetic interaction via Si tetrahedron can be attributed to the length of the O–O edges of 2.748 Å in the tetrahedron, which are longer than the O–O edges of 2.704 Å in Al octahedron. Hence, two magnetic sublattices connected and coupled antiferromagnetically (Fig. 5.15) via Si tetrahedra and Al octahedra occur in the structure of almandine.

Table 5.7 Calculated spin-coupling constants and quadrupole splittings between six iron ions connected via Si tetrahedra and Al octahedra

Connecting polyhedron	N_{at}	J, cm^{-1}	Magnetic configuration	ΔE_Q, mm·s^{-1}					
				1	2	3	4	5	6
Si	165	-0.07	FM	-3.47	-3.56	-3.56	-3.47	-3.56	-3.56
			AM	-3.48	-3.56	-3.56	-3.48	-3.56	-3.56
Al	165	-0.78	FM	-3.46	-3.46	-3.46	-3.46	-3.46	-3.46
			AM	-3.47	-3.47	-3.47	-3.47	-3.47	-3.47

Such a spin structure is in accordance with previous experimental investigations on almandine, where T_N of 7.5 K is in accordance with the calculated weak antiferromagnetic interaction. Generally small values of the spin coupling constant are associated with relatively large Fe–O distances in almandine.

Finally, each magnetic sublattice produces its own hyperfine pattern in the Mössbauer spectrum. If the magnetic sublattices are equal, then the Mössbauer parameters, i.e. intensity, quadrupole splitting, etc. have to be identical, as it is. Therefore, the derived antiferromagnetic interaction of two identical magnetic sublattices perfectly explains the

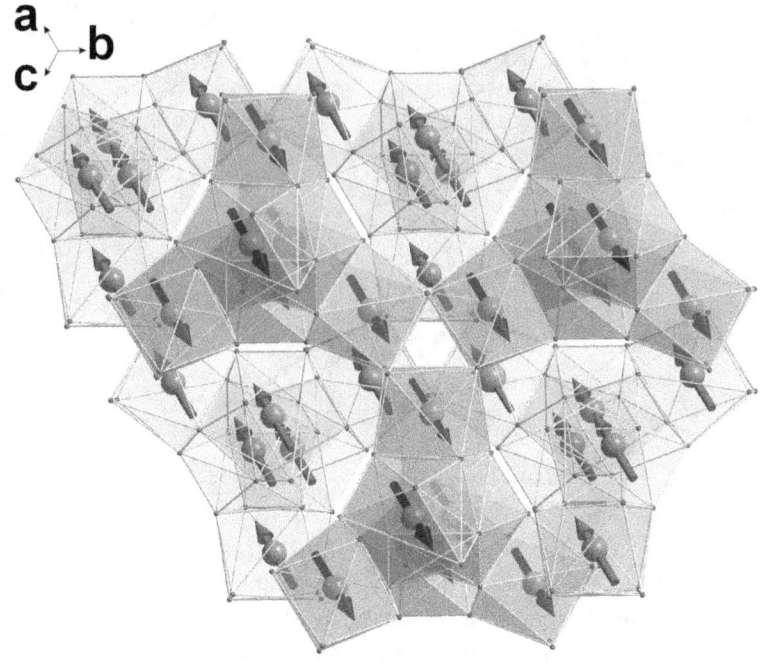

Figure 5.15 (111) projection of calculated spin pattern in almandine:
two magnetic sublattices.

complex spectrum of two nearly identical eight-line magnetic hyperfine patterns (de Oliveira et al. 1987 and 1989, Murad and Wagner 1987).

5.4 Summary

The electronic and magnetic structure of almandine has been characterized by electronic structure calculations in the local spin density approximation. Whereas the d-d energies are somewhat smaller than the respective experimental values, they are still in reasonable agreement. The calculated magnetic moment, isomer shift and quadrupole splittings are in quantitative agreement with the corresponding experimental values, allowing the conclusion that the structure of almandine is correctly being described, at least qualitatively. The almandine structure contains two identical sublattices of Fe dodecahedra connected via Al octahedra and Si tetrahedra. The magnetic interaction between the iron spins via oxygen bridges within each sublattice is ferromagnetic, whereas, the coupling between two magnetic sublattices is weakly antiferromagnetic via Al octahedra and Si tetrahedra. This antiferromagnetic interaction of two identical magnetic sublattices is in qualitative agreement with experiment (Prandl 1971, Anovitz et al. 1993), and explains and solves controversy in the interpretation of the Mössbauer spectra of almandine below the Néel temperature.

Chapter 6

ELECTRONIC AND MAGNETIC STRUCTURE OF ANDRADITE

6.1 Introduction

In this chapter we investigate another representative Fe-bearing silicate by the SCC-Xα method, viz. the Fe^{3+} bearing garnet andradite, $Ca_3Fe^{3+}{}_2Si_3O_{12}$, which is a rock-forming mineral typically found in skarns and contact metamorphic assemblages. The general chemical formula of garnets is $\{X_3\}[Y_2](Z_3)O_{12}$, where $\{\}$, $[\,]$ and $(\,)$ denote dodecahedral, octahedral and tetrahedral sites, respectively, following Geller's notation (1967). As the other garnets andradite crystallizes in the cubic space group $Ia\bar{3}d$ (Fig. 6.1), with unit-cell parameter $a = 12.051$ Å at 100 K (Armbruster T. and Geiger C.1993). The fractional atomic coordinates are presented in Table 6.1 and the respective coordinates of equivalent positions for the general site are listed in Table 6.2.

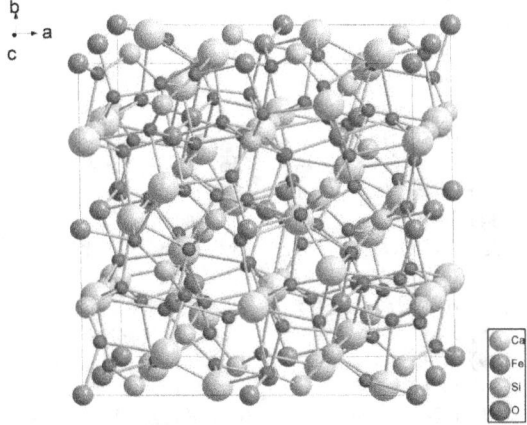

Figure 6.1 Unit cell of andradite.

The structure of andradite is built up by the Fe^{3+} cations, which are coordinated to 6 oxygens in almost regular octahedral arrangement (Y-site), the Ca cations which are dodecahedrally coordinated to 8 oxygens (X-site), and the Z-sites coordinating to 4 oxygens in a tetrahedral arrangement which are occupied exclusively by Si atoms in silicate garnets. The Fe octahedra have Fe–O bond lengths of 2.019 Å and the O–Fe–O angles of 89.2° and 90.8° are close to the ideal value of 90° (Fig. 6.2 a). Oxygen atoms at the opposite corners of

93

the octahedron form O–Fe–O angles of 180^0. Therefore, the octahedra are almost ideal and can be described as having the symmetry C_3 (Deer et al 1997), C_{3i} (Manning and Townsend 1969, Gaft et al. 2005) or even S_6 (Novak and Gibbs 1971). However, the consideration of the geometry of the octahedron (Fig.6.2a) reveals the D_{3d} symmetry in our case. In addition, it was assumed so far that the Al octahedron in almandine has S_6 symmetry (Novak and Gibbs 1971), however, it is trigonally elongated octahedron with D_{3d} symmetry too. The Si tetrahedra can be described by the C_4 symmetry (Deer et al 1997, Novak and Gibbs 1971) with the Si-O bond lengths of 1.649 Å and O–Si–O angles of 102.3^o and 113.2^o (Fig. 6.2 b). In turn, the 8-fold coordinated polyhedron around the Ca^{2+} cation can be described as distorted triangular dodecahedron with D_2 symmetry (Deer et al 1997, Novak and Gibbs 1971) and with the Ca–O bond lengths of 2.358 Å and 2.495 Å (Fig. 6.2 c).

Table 6.1 Fractional atomic coordinates for synthetic andradite at 100 K (Sueno 1976)

Atom	Oxidation	x	y	z
Fe	+3	0.00000	0.00000	0.00000
Ca	+2	0.00000	0.25000	0.12500
Si	+4	0.00000	0.25000	0.37500
O	–2	0.03914(6)	0.04895(6)	0.65534(6)

Table 6.2 Equivalent positions for the space group $Ia\bar{3}d$, No. 230 (Hahn 1987)

(1) x,y,z	(25) 3/4-y,1/4+x,3/4+z	(49) 1/2-x,-y,1/2+z	(73) 3/4+y,3/4-x,1/4+z
(2) 1/2+x,1/2+y,1/2+z	(26) 1/4-y,3/4+x,1/4+z	(50) -x,1/2-y,+z	(74) 1/4+y,1/4-x,3/4+z
(3) z,1/2-x,1/2+y	(27) 1/4+x,1/4+z,1/4+y	(51) 1/2-z,1/2+x,y	(75) 1/4-x,3/4-z,3/4+y
(4) 1/2+z,-x,+y	(28) 3/4+x,3/4+z,3/4+y	(52) -z,+x,1/2+y	(76) 3/4-x,1/4-z,1/4+y
(5) -y,1/2+z,1/2-x	(29) 1/4-z,1/4-y,1/4-x	(53) 1/2+y,1/2-z,-x	(77) 1/4+z,3/4+y,3/4-x
(6) 1/2-y,+z,-x	(30) 3/4-z,3/4-y,3/4-x	(54) +y,-z,1/2-x	(78) 3/4+z,1/4+y,1/4-x
(7) x,1/2-y,1/2+z	(31) 1/4+y,1/4+x,1/4+z	(55) 1/2-x,1/2+y,z	(79) 1/4-y,3/4-x,3/4+z
(8) 1/2+x,-y,+z	(32) 3/4+y,3/4+x,3/4+z	(56) -x,+y,1/2+z	(80) 3/4-y,1/4-x,1/4+z
(9) -z,1/2+x,1/2-y	(33) 1/4-x,1/4-z,1/4-y	(57) 1/2+z,1/2-x,-y	(81) 1/4+x,3/4+z,3/4-y
(10) 1/2-z,+x,-y	(34) 3/4-x,3/4-z,3/4-y	(58) +z,-x,1/2-y	(82) 3/4+x,1/4+z,1/4-y
(11) y,1/2-z,1/2+x	(35) 1/4+z,1/4+y,1/4+x	(59) 1/2-y,1/2+z,x	(83) 1/4-z,3/4-y,3/4+x
(12) 1/2+y,-z,+x	(36) 3/4+z,3/4+y,3/4+x	(60) -y,+z,1/2+x	(84) 3/4-z,1/4-y,1/4+x
(13) -x,1/2+y,1/2-z	(37) 1/4-y,1/4-x,1/4-z	(61) 1/2+x,1/2-y,-z	(85) 1/4+y,3/4+x,3/4-z
(14) 1/2-x,+y,-z	(38) 3/4-y,3/4-x,3/4-z	(62) x,-y,1/2-z	(86) 3/4+y,1/4+x,1/4-z
(15) 3/4+x,1/4-z,3/4-y	(39) 1/2+z,x,1/2-y	(63) 3/4-x,3/4+z,1/4-y	(87) -z,-x,-y
(16) 1/4+x,3/4-z,1/4-y	(40) +z,1/2+x,-y	(64) 1/4-x,1/4+z,3/4-y	(88) 1/2-z,1/2-x,1/2-y
(17) 3/4-z,1/4+y,3/4+x	(41) 1/2-y,-z,1/2+x	(65) 3/4+z,3/4-y,1/4+x	(89) y,z,x
(18) 1/4-z,3/4+y,1/4+x	(42) -y,1/2-z,+x	(66) 1/4+z,1/4-y,3/4+x	(90) 1/2+y,1/2+z,1/2+x
(19) 3/4+y,1/4-x,3/4-z	(43) 1/2+x,y,1/2-z	(67) 3/4-y,3/4+x,1/4-z	(91) -x,-y,-z
(20) 1/4+y,3/4-x,1/4-z	(44) +x,1/2+y,-z	(68) 1/4-y,1/4+x,3/4-z	(92) 1/2-x,1/2-y,1/2-z
(21) 3/4-x,1/4+z,3/4+y	(45) 1/2-z,-x,1/2+y	(69) 3/4+x,3/4-z,1/4+y	(93) z,x,y
(22) 1/4-x,3/4+z,1/4+y	(46) -z,1/2-x,+y	(70) 1/4+x,1/4-z,3/4+y	(94) 1/2+z,1/2+x,1/2+y
(23) 3/4+z,1/4-y,3/4-x	(47) 1/2+y,z,1/2-x	(71) 3/4-z,3/4+y,1/4-x	(95) -y,-z,-x
(24) 1/4+z,3/4-y,1/4-x	(48) +y,1/2+z,-x	(72) 1/4-z,1/4+y,3/4-x	(96) 1/2-y,1/2-z,1/2-x

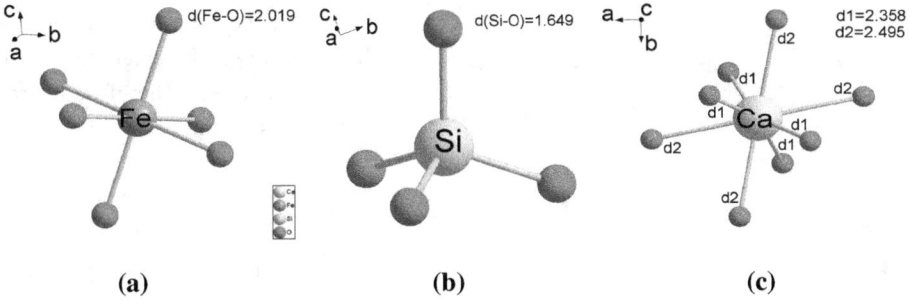

Figure 6.2 Cation polyhedra in the structure of andradite: **(a)** – Fe^{3+} octahedron, **(b)** - Si tetrahedron, **(c)** - Ca dodecahedron.

Every oxygen atom is bonded to one Si, one Fe^{3+} and two Ca atoms. In contrast to orthoferrosilite and almandine, the iron polyhedra of andradite do not share corners or edges with one another, i.e. they are isolated from each other. Accordingly, iron octahedra are connected only via Si tetrahedra and/or Ca dodecahedra with a nearest Fe–Fe distance of 5.22 Å. In turn, every Fe octahedron has in its direct environment six corner-shared Si tetrahedra and six edge-shared Ca dodecahedra (Fig. 6.3).

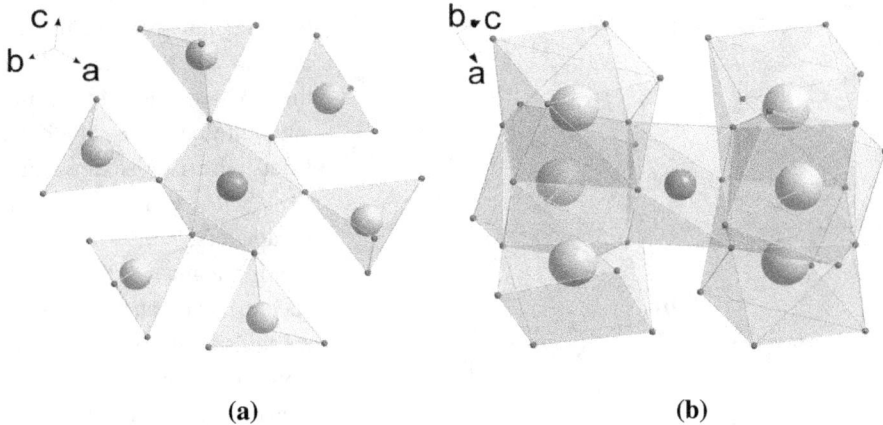

Figure 6.3 Direct environment of Fe^{3+} octahedron: **(a)** – six corner-shared Si tetrahedra, **(b)** – six edge-shared Ca dodecahedra.

As described in Chapter 3, Mössbauer and absorption spectroscopic measurements provide the best opportunity to look at local characteristics of a structure. Therefore, andradite was subject of numerous investigations by optical spectroscopy (Manning et al. 1967, 1968, 1969 Parkin and Burns 1980, Qing-Han 1987, Taran and Langer 2000) and Mössbauer spectroscopy (Amthauer 1976, Murad 1984, Armbruster and Geiger 1993, Woodland and Ross 1994, Manning and Tricker 1977, Kühberger et al. 1989).

Since iron ions in silicates are always in the high-spin state, andradite should exhibit a magnetic phase transition from a paramagnetic to a magnetic state at low temperatures. Actually, the Mössbauer spectrum of andradite below T_N=11 K can be fitted with two sextets (Fig. 6.4 a) with nearly identical parameters; this is the most striking feature of the Mössbauer spectra, since andradite structure contains only a single crystallographic Fe^{3+} site. A conceivable interpretation of the existence of two sextets is that the sublattice magnetization and the EFG are oriented along different directions. Further heating (Murad 1984) of andradite reveals that antiferromagnetic ordering begins to disappear at a Néel temperature of 11.4 K (Fig. 6.4 b). For temperatures above T_N the spectrum consist of a paramagnetic doublet with average quadrupole splitting ΔE_Q of 0.55 mm·s^{-1} at 121 K (Murad

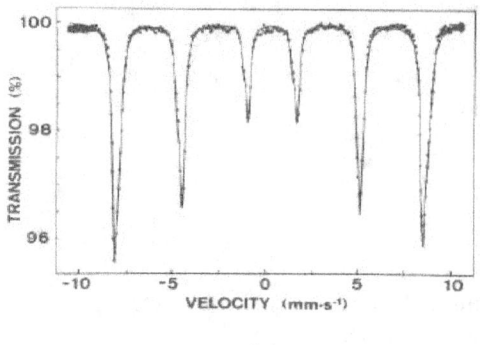

(a)

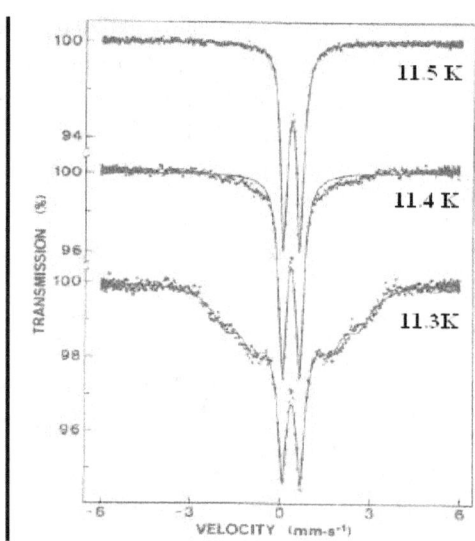

(b)

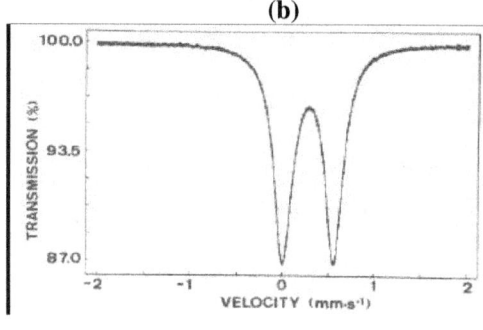

(c)

Figure 6.4 Mössbauer spectra of (Murad 1984):
(**a**) – antiferromagnetically ordered andradite (4.2 K),
(**b**) – andradite in the vicinity of the Néel temperature,
(**c**) – paramagnetic andradite (121 K).

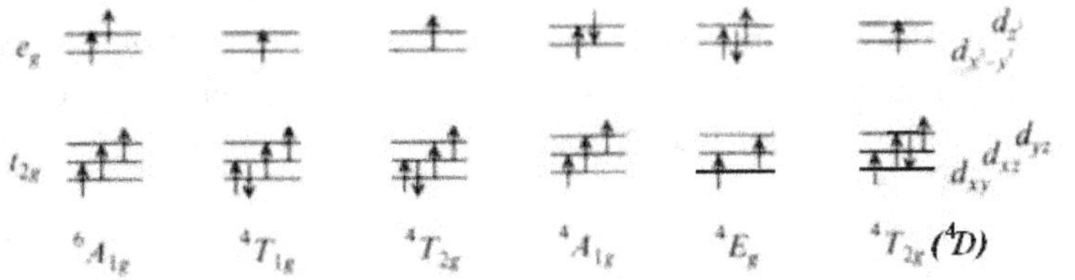

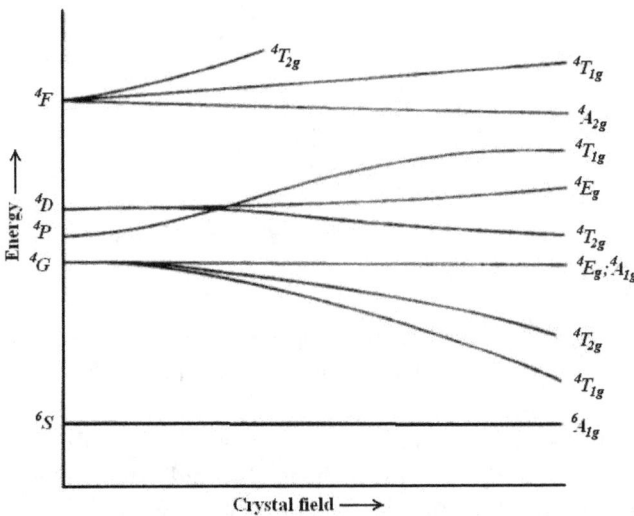

Figure 6.5 Electronic configurations of Fe^{3+} in octahedral coordination for the ground state and some low-lying quartet states.

1984) and isomer shift δ of 0.39 mm·s^{-1} (Fig. 6.4 c). There is almost no difference between the values of the quadrupole splitting at RT and LNT, whereas the isomer shift is sensitive to the temperature changes and has values of 0.40 and 0.49 mm·s^{-1} at RT and LNT (Armbruster and Geiger 1993), respectively.

Since the electronic ground state of Fe^{3+} in andradite is $^6A_{1g}$ (notation in ideal octahedron) corresponding to high-spin electronic configuration $3d^5$ with one unpaired electron in each of the three t_{2g} and the two e_g orbitals, spin-allowed d–d transitions are not possible due to the Pauli principle (Section 3.3). Accordingly, four weak absorption bands at 10640, 16040, 23020 and 25540 cm^{-1} (Taran and Langer 2000) are assigned to the spin-forbidden electronic d-d transitions from the sextet ground state, $^6A_{1g}$, to the quartet excited states, $^4T_{1g}$, $^4T_{2g}$, $^4A_{1g}$, ^4E_g (arising from ^4G) and $^4T_{2g}$ (arising from ^4D) (Fig. 6.5), respectively. The relative energies of the crystal field states of octahedrally coordinated ferric iron may also be illustrated by a Tanabe-Sugano diagram (Tanabe and Sugano 1954):

Figure 6.6 Tanabe-Sugano diagram (after Burns 1993) for Fe^{3+} in octahedral coordination.

The purpose of the subsequent theoretical investigations of andradite is the comprehensive characterization of its electronic and magnetic structure. As usual, the reliability and suitability of the model clusters will be assessed by the comparing of the calculated spectroscopic data with the experimental results.

6.2 Electronic Structure of Andradite

The electronic structure calculations are based on the single crystal structural data determined by X-ray diffraction measurements on well characterized synthetic end-member andradite at 100 K (Armbruster and Geiger 1993). Unlikely orthoferrosilite (Chapter 4) and almandine (Chapter 5), andradite contains Fe^{3+} ions. Hence, iron ions in more distant octahedra have to be substituted by trivalent main group metal ions as Al^{3+} ions. Consequently, the valence basis set consists of 2s-, 2p-orbitals for O and F, 3s-, 3p-orbitals for Si and Al, 4s-, 4p-orbitals for Ca and 3d-, 4s-, 4p-orbitals for Fe.

The cations of the second coordination sphere around an $Fe^{3+}O_6$ octahedron comprise six tetravalent Si ions and six divalent Ca cations. Adding these twelve polyhedra yields the cluster $[Fe^{3+}O_6\text{-}Ca_6Si_6O_{36}]^{45-}$ with 36 oxygen atoms in the third shell (Fig. 6.7). Therefore,

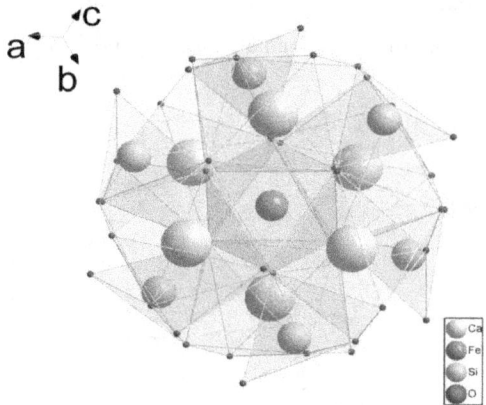

Figure 6.7 The 1, 2 and 3 coordination spheres of the central Fe atom within the crystal structure of andradite.

some of the oxygen atoms have to be replaced with fluorine atoms or omitted at all (Sections 3.4). Consequently, the resulting cluster containing all coordination polyhedra of the cations bonded to the oxygen atoms of the first coordination sphere of the central iron ion comprises 43 atoms and has the composition $[Fe^{3+}O_6\text{-}Si_6Ca_6F_{24}]^{3+}$.

Table 6.3 Constructed clusters of increasing size comprise N_{at} atoms, $N_{v.o.}$ valence orbitals and $N_{v.e.}$ valence electrons resulting the positive cluster charge

Composition	N_{at}	$N_{v.o.}$	$N_{v.e.}$
$[Fe^{3+}O_6]^{9-}$	7	33	53
$[Fe^{3+}O_6\text{-}Si_6F_{18}]^{3-}$	31	129	197
$[Fe^{3+}O_6\text{-}Si_6Ca_6F_{24}]^{3+}$	43	177	245
$[Fe^{3+}O_6\text{-}Si_6Ca_6Al_6F_{42}]^{3+}$	67	273	389
$[Fe^{3+}O_6\text{-}Si_{12}Ca_6Al_8O_6F_{60}]^{3+}$	99	401	581
$[Fe^{3+}O_6\text{-}Si_{18}Ca_6Al_8O_{18}F_{60}]^{3+}$	117	473	677
$[Fe^{3+}O_6\text{-}Si_{18}Ca_{12}Al_8O_{24}F_{54}]^{3+}$	129	521	725
$[Fe^{3+}O_6\text{-}Si_{18}Ca_{12}Al_{14}O_{30}F_{66}]^{3+}$	147	593	821
$[Fe^{3+}O_6\text{-}Si_{18}Ca_{18}Al_{14}O_{36}F_{66}]^{3+}$	159	641	869
$[Fe^{3+}O_6\text{-}Si_{18}Ca_{24}Al_{14}O_{36}F_{78}]^{3+}$	177	713	965

The subsequent stepwise extension of the cluster size is realized by adding coordination polyhedra with simultaneously replacing additional iron ions by aluminium ions and incompletely coordinated oxygen ions with fluorine ions, respectively, or omitting the latter. This extension of the cluster size has been accompanied by preserving the local symmetry of central Fe^{3+} ion and the total charge of the cluster (i.e., $Q= +3$). Such a strategy has led to a hierarchy of clusters of increasing size (Table 6.3) that may serve as a control for size convergence. The largest size-converged cluster constructed in this way and used for the calculation of the spectroscopic data of iron contains 177 atoms (Table 6.3). Moreover, a

Table 6.4 Constructed clusters of increasing size comprise N_{at} atoms, $N_{v.o.}$ valence orbitals and $N_{v.e.}$ valence electrons resulting the negative cluster charge

Composition	N_{at}	$N_{v.o.}$	$N_{v.e.}$
$[Fe^{3+}O_6]^{9-}$	7	33	53
$[Fe^{3+}O_6\text{-}Si_6F_{18}]^{3-}$	31	129	197
$[Fe^{3+}O_6\text{-}Si_6Ca_6F_{24}]^{3+}$	43	177	245
$[Fe^{3+}O_6\text{-}Si_6Ca_6Al_6F_{42}]^{3+}$	97	393	533
$[Fe^{3+}O_6\text{-}Si_{12}Ca_{12}Al_{12}O_{24}F_{54}]^{3-}$	121	489	677
$[Fe^{3+}O_6\text{-}Si_{12}Ca_{18}Al_{14}O_{30}F_{60}]^{3-}$	141	569	773
$[Fe^{3+}O_6\text{-}Si_{18}Ca_{18}Al_{14}O_{42}F_{60}]^{3-}$	159	641	869
$[Fe^{3+}O_6\text{-}Si_{18}Ca_{24}Al_{20}O_{54}F_{66}]^{3-}$	189	761	1013

series of model constructions of the clusters leading to the negative total charge of −3 (Table 6.4) is realized in order to analyze the influence of the sign of the total cluster charge on the orbital energy levels, quadrupole splittings and isomer shifts. In this case, the largest size-converged cluster used for the calculation of the spectroscopic data of iron contains 189 atoms (Table 6.4). The clusters with the negative charge are compensated by the Madelung potential, whereas positively charged clusters do not contain this compensation.

Both largest clusters of 177 and 189 atoms with the total charges of +3 and −3, respectively, comprise all atoms around the central iron atom within the fifth coordination sphere of 7.82 Å (Fig. 6.8), and all anions are correctly represented as oxygen atoms within a sphere of 5.89 Å.

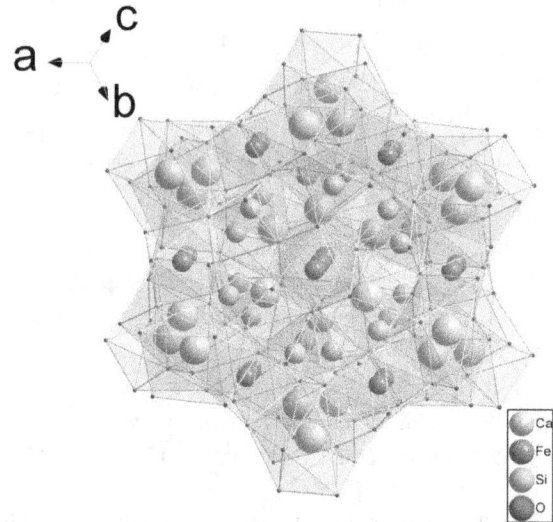

Figure 6.8 The environment of central Fe^{3+} ion within 1-5 coordination spheres is described by the largest cluster of 177 atoms.

The results for the quadrupole splitting and energy levels of the spin-down *3d* orbitals are given in Tables 6.5 and 6.6. It is seen (Table 6.5) that increasing the cluster size from 117 up to 177 atoms does not change appreciably the 3d-energies or quadrupole splitting, so that a cluster of 117 atoms appears to be size convergent with respect to the calculated spectroscopic data. In the case of the series leading to the negative cluster charge, the cluster of 121 atoms is size convergent.

Since spin-down *3d* orbitals are not occupied the optical absorption bands are caused by spin-forbidden transitions (Section 6.1) which cannot reliably be described by the SCC-Xα method. Therefore, only the calculated hyperfine parameters can be used for comparison with experiments (Table 6.7). Nevertheless, in accordance with qualitative estimation the crystal-field splitting of spin-down orbitals must be below about 3eV since otherwise Fe^{3+} would be in the low-spin state.

Table 6.5 Calculated energies (in cm^{-1}) of $3d$-like orbitals with respect to the lowest 3d-orbital (ε_0) and quadrupole splitting ΔE_Q (in mm·s^{-1} with η in brackets) for andradite clusters of increasing size with charge of q around Fe^{3+} ion resulting the positively charged size-converged cluster

Number of atoms	q	$\varepsilon_1(\pi^*)$	$\varepsilon_2(\pi^*)$	$\varepsilon_3(\sigma^*)$	$\varepsilon_4(\sigma^*)$	$\Delta E_Q(\eta)$
7	-9	64	64	10768	10768	0.143(0.00)
31	-3	428	428	13382	13382	0.311(0.00)
43	+3	694	694	26795	26795	0.634(0.00)
67	+3	508	508	28417	28417	0.301(0.00)
99	+3	1033	1033	25860	25860	0.486(0.00)
117	+3	960	960	26851	26851	0.467(0.00)
129	+3	951	951	28069	28069	0.505(0.00)
147	+3	1226	1226	28481	28481	0.698(0.00)
159	+3	1218	1218	29828	29828	0.527(0.00)
177	+3	1226	1226	29021	29021	0.559(0.00)

Table 6.6 Calculated energies (in cm^{-1}) of $3d$-like orbitals with respect to the lowest 3d-orbital (ε_0) and quadrupole splitting ΔE_Q (in mm·s^{-1} with η in brackets) for andradite clusters of increasing size with charge of q around Fe^{3+} ion resulting the negatively charged size-converged cluster

Number of atoms	q	$\varepsilon_1(\pi^*)$	$\varepsilon_2(\pi^*)$	$\varepsilon_3(\sigma^*)$	$\varepsilon_4(\sigma^*)$	$\Delta E_Q(\eta)$
7	-9	64	64	10768	10768	0.143(0.00)
31	-3	428	428	13382	13382	0.311(0.00)
43	+3	694	694	26795	26795	0.634(0.00)
97	+3	807	807	20004	20004	0.453(0.00)
121	-3	1048	1048	15566	15566	0.634(0.00)
141	-3	1016	1016	14333	14333	0.685(0.00)
159	-3	1008	1008	14705	14705	0.689(0.00)
189	-3	1097	1097	11994	11994	0.732(0.00)

Accordingly, the energies of the e_g-like orbitals are overestimated for the clusters with positive charge (Table 6.5) and in the case of the negatively charged clusters with the Madelung potential compensation, the e_g-like orbitals have reasonable values (Table 6.6). From the calculated energies of $3d$ orbitals, it is seen that the t_{2g}-like orbitals are split into two levels, whereas the e_g-like orbitals remain degenerate. The splitting of the t_{2g}-like orbitals reflects the distortion of iron octahedra from ideal symmetry arising from the deviation of 0.8° (less than 1%) of the O–Fe–O angles from 90° (Section 6.1). Moreover, the degeneracy of the e_g orbitals indicates that the Fe^{3+} polyhedron is the trigonally compressed octahedron (Section 3.1) at 100 K. In addition, the electric field gradient is axially symmetric, i.e. $\eta = 0.0$ for all clusters demonstrating the conservation of the symmetry of the model clusters and reflecting axial symmetry around the central iron ion.

The calculated isomer shift δ of 0.59 mm·s^{-1} for the positively charged cluster is overestimated too (Tab. 6.7), while the isomer shift of 0.45 mm·s^{-1} for the negatively charged cluster is in agreement with the experimental values within the expected error margins of the calculations. However, the calculated quadrupole splitting of 0.56 mm·s^{-1} for the 177-atom cluster (Q=+3) is in quantitative agreement with the experimental values between 0.55 and 0.60 mm·s^{-1}, whereas the quadrupole splitting of 0.73 mm·s^{-1} for the 189-atom cluster (Q= −3) is overestimated (Table 6.7).

Table 6.7 Comparison of the calculated and measured quadrupole splittings ΔE_Q and isomer shifts δ (in mm·s^{-1}) for size-converged 177 and 189 atom clusters around the Fe^{3+} cation in andradite with charges of +3 and -3, respectively

	calc.[1]	calc.[2]	exp[3]	exp[4]	exp[5]	exp[6]	exp[7]
ΔE_Q	0.56	0.73	0.55	0.55	0.56	0.55	0.60
δ	0.59	0.45	0.50	0.39	0.49	0.48	0.39

[1] Theoretical calculations for 177 atom cluster with total charge of +3.
[2] Theoretical calculations for 189 atom cluster with total charge of −3.
[3] Amthauer G. et al. (1976) Z. Kristallographie 143, p. 14–55, (T=77K).
[4] Murad E. (1984) Am. Mineral. 69, p. 722–724, (T=121K).
[5] Armbruster T. and Geiger C. (1993) Eur. J. Mineral 5, p. 59-71, (T=77K).
[6] Woodland A. and Ross C. II (1994) Phys. Chem. Minerals. 21, p. 117–132, (T=80K).
[7] Manning P. and Tricker M. (1977) Canad. Miner. 15, p. 81-86, (T=300K).

This difference in the calculated hyperfine parameters can be explained by the larger number of the singly-bonded fluorine atoms from 12 to 18 for clusters with charges of +3 and −3, respectively. As a result the covalent bonding character decreases, and the total occupation number of 4s and 3d orbitals increases leading to the decreasing the screening of s electrons in the inner shells. In result, s electron densities increase (Table 6.8), and therefore the isomer shift decreases. In turn, from Table 6.9 we can see that the d anisotropy (equation 3.24) increases from 0.08 to 0.12 in going from positively to negatively charged clusters. Consequently, the quadrupole splitting is larger for the negatively charged cluster containing more fluorine atoms.

Table 6.8 Electron density contributions of s-orbital to the isomer shift

$N_{at.}(Q)$	Occ.num. (4s+3d)	electron density contribution				$\Delta\rho(0)$
		4s	3s	2s	1s	
177(+3)	6.507	1.405	2.926	-2.359	0.655	-2.110
189(-3)	6.768	1.736	3.046	-2.294	0.626	-1.624

Table 6.9 Orbital occupation numbers in the axes system of the EFG for the largest positively and negatively charged clusters

$N_{at.}(Q)$	n_s	n_x	n_y	n_z	n_{xy}	n_{yz}	n_{z^2}	n_{xz}	$n_{x^2-y^2}$
177(+3)	0.669	0.258	0.230	0.258	1.176	1.215	1.057	1.215	1.176
189(-3)	0.757	0.134	0.115	0.134	1.215	1.243	1.096	1.243	1.215

The calculated effective charges Q_{eff} (Table 6.10) of Fe^{3+} and O ions for positively and negatively charged clusters exhibit substantial deviations from the formal oxidation states of +2 and –2, respectively. These deviations indicate considerable covalent character of the iron-oxygen bonds within the structure of andradite. A qualitative measure for the covalent part of a bond is the overlap population. The values listed in Table 6.10 show that more than 85% of the total overlap populations arise from the orbital interactions of the (formally empty) iron 4s, 4p orbitals with the valence orbitals of the oxygens, whereas the completely filled 3d↑ orbitals yield a small negative (antibonding) contribution because both the bonding and the antibonding linear combinations with the oxygen orbitals are occupied. The small contributions from the 3d↓ orbitals arise from the admixtures of the five formally empty 3d↓ orbitals to the bonding spin-down molecular orbitals, whereas the occupied 3d↑ -orbital has almost pure 3d-character and does not contribute to the Fe-O bonds. Again, it is seen from Table 6.10 that the covalent character of the Fe–O bonds decreases in going from positively to negatively charged cluster. Furthermore, the valence shell occupations of Fe^{3+} contain considerable amounts from 4s and 4p orbitals, $4s^{0.67}4p^{0.75}3d^{5.84}$ and $4s^{0.76}4p^{0.38}3d^{6.01}$ for positively and negatively charged clusters, respectively. The increased 3d occupation arises from the admixture of the five formally empty 3d↓ orbitals to the bonding spin-down molecular orbitals. Consequently, the 3d spin densities are reduced from the free-ion value 5 to 4.21 and 4.03 for clusters with charges of +3 and –3, respectively.

Table 6.10 Effective charges and contributions to the overlap populations in the iron-oxygen bonds

$N_{at.}(Q)$	$Q_{eff}(Fe)$	$Q_{eff}(O)$	Total	$4s + 4p$	3d↑	3d↓
177(+3)	+0.75	–0.42	0.224	0.196	-0.015	0.043
189(–3)	+0.85	–0.56	0.204	0.174	-0.015	0.044

The calculated effective magnetic moments in Fe^{3+} ions of 5.11 and 4.93 μ_B for clusters with charges of +3 and –3, respectively, are in quantitative agreement with the experimental value of 5.22 μ_B (Manning and Harris 1970) within the expected error margins of the calculations. This is below the 5.92·μ_B found for magnetically isolated Fe^{3+} ion in the S=5/2 state due to the reduction of the spin density of Fe^{3+} cation caused by covalence interaction with surrounding anions and showing low extent of the magnetic interaction.

6.3 Magnetic Structure of Andradite

In order to explain the spin structure of andradite the possible interaction pathways have to be carefully considered. The visualization tool for the crystal structure data, Diamond 3 (http://www.crystalimpact.com/diamond/Default.htm), was used for this purpose. Thereupon, the corresponding spin coupling constants between spins of Fe^{3+} ions have to be calculated for a series of clusters of increasing size. Since crystal structure data of andradite for the temperatures below Néel temperature are not available, all calculations are performed for the structure at 100 K (Armbruster and Geiger 1993).

Since each Fe octahedron in andradite does not share any edge or corner with other Fe octahedra, the spins of Fe^{3+} ions only interact via edges of Si tetrahedra and Ca dodecahedra. As mentioned in Section 6.1, the direct environment of every Fe octahedron consists of 6 corner-shared Si-tetrahedra and 6 edge-shared Ca-dodecahedra (Fig. 6.3).

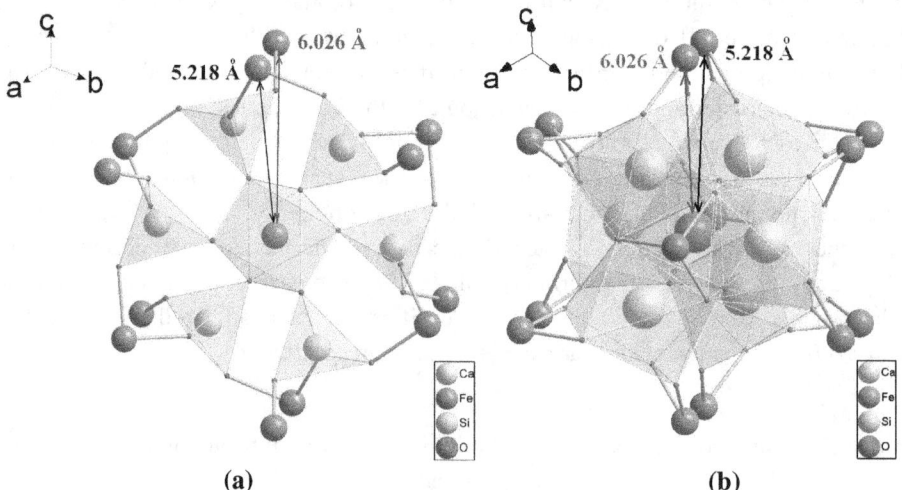

Figure 6.9 Possible connections between iron atoms via: **(a)** – Si tetrahedra, **(b)** – Ca dodecahedra.

Firstly, let us consider the connection between Fe octahedra via Si tetrahedra. There are two types of terminal iron ions. The terminal iron cations of the first type are connected to two Si tetrahedra with distances of 5.218 Å to the central Fe ion. The iron cations of the second type are connected to one Si tetrahedron with distances of 6.026 Å to the central Fe ion (Fig. 6.9 a).

Secondly, in the case of the connection via Ca dodecahedra, there are three types of terminal iron ions, which are connected to two Ca dodecahedra with distances of 5.218 and 6.026 Å to the central Fe ion (Fig. 6.9 b). Terminal iron ions of the third type are connected to three Ca dodecahedra and situated in front of and behind the central iron ion along the (111) direction at the distance of 5.218 Å (Fig. 6.9 b).

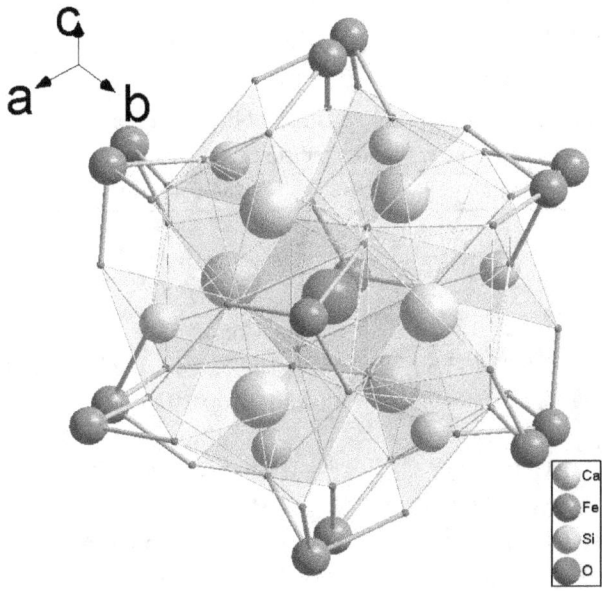

Figure 6.10 All possible connections between neighboring iron atoms within the structure of andradite.

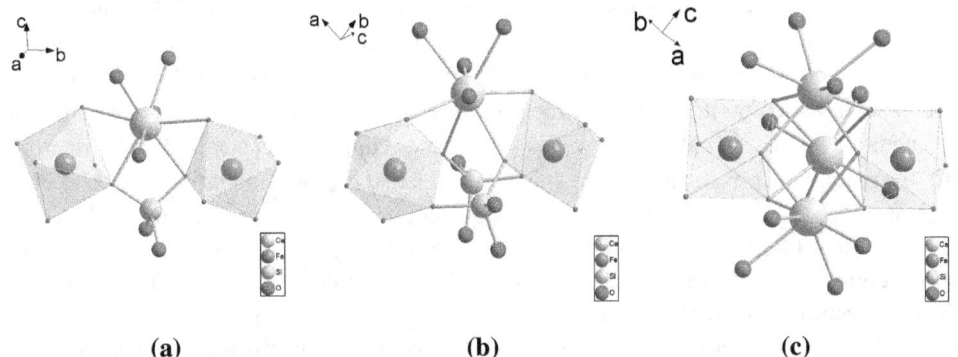

(a)   (b)   (c)

Figure 6.11 Possible interaction pathways in the structure of Andradite via edges of:
(a) – one Ca and one Si polyhedra with d(Fe-Fe)=6.026 Å,
(b) – one Ca dodecahedron and two Si tetrahedra with d(Fe-Fe)=5.218 Å,
(c) – three Ca dodecahedra with d(Fe-Fe)=5.218 Å.

If we combine both possible connection groups via Ca and Si polyhedra, we obtain 3 types of connections (Fig. 6.10). The first one is via one Ca and one Si polyhedra at the distance of 6.026 Å to the central iron ion (Fig. 6.11 a), the second one is via one Ca dodecahedron and 2 Si tetrahedra at the distance of 5.218 Å (Fig. 6.11 b) and the third is via three Ca dodecahedra along (111) direction at the distance of 5.218 Å (Fig. 6.11 c). Using these possible exchange pathways, three neutral clusters with increasing size which describe the magnetic interaction between the iron ions are constructed (Section 6.2) and respective spin-coupling constants were calculated (Table 6.11). From the Table 6.11 one can see that the values of the quadrupole splittings are overestimated due to the initial negatively charged cluster used for the modelling of the magnetic interaction.

Table 6.11 Calculated spin coupling constants J between neighboured Fe^{3+} with some geometrical and spectroscopic parameters

	Fe–Si,Ca–Fe	Fe–2Si,Ca–Fe	Fe–3Ca–Fe
J, cm^{-1}	−1.9	+1.4	+0.6
d(Fe–Fe), Å	6.026	5.218	5.218
d(O–O), Å	2.569	Si: 2.753	2.926
		Ca: 2.855	
$\angle$(Fe,O,O)	133.59°, 133.59°	Si: 100.22°, 144,78°	124.43°, 124,43°
		Ca: 97.00°, 97.00°	
N_{at}	243	221	205
ΔE_Q, mm·s^{-1}	FM: 0.63, 0.63	FM: 0.64, 0.64	FM: 0.69, 0.69
	AM: 0.63, 0.63	AM: 0.63, 0.63	AM: 0.69, 0.69
δ, mm·s^{-1}	FM: 0.51, 0.51	FM: 0.50, 0.50	FM: 0.52, 0.52
	AM: 0.51, 0.51	AM: 0.50, 0.50	AM: 0.52, 0.52

The small values of the calculated spin coupling constants reflect the weak magnetic interaction between the spins of the iron ions, which is consistent with the low value of 11.4K for the Néel temperature. The spin coupling constant of negative sign is the largest reflecting the predominant antiferromagnetic interaction which is in agreement with the experiments. Thereupon, the attempts to figure out the spin orientation in the structure of andradite have been made based on the calculated values of the spin coupling constants. It was ascertained that the structure is spin frustrated and spins are canted. The spin frustration can be explained if we consider a square which is formed by the iron ions with the separation of 6.026 Å (Fig. 6.12). These four ions have one common nearest neighboring iron ion that is connected to those by equivalent bridges with Fe–Fe distance of 5.218 Å. The spins of the four corner iron ions are coupled antiferromagnetically, whereas each of them is coupled ferromagnetically with the central one. From the Fig. 6.12 one can see that these two requirements cannot be simultaneously satisfied, and therefore the central spin (Fe5) is canted. Due to the equivalence of the bridges with four corner irons it is possible to suggest that the canting angle is 90°.

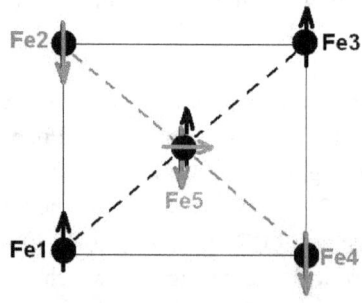

Figure 6.12 The reason of the spin frustration.

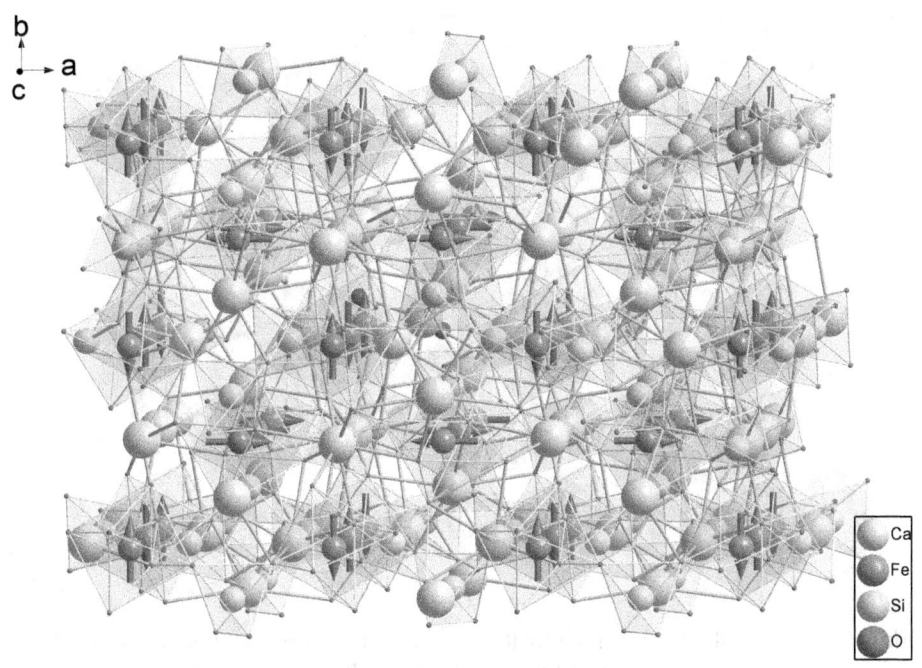

Figure 6.13 The 3D spin pattern in andradite: spin frustration.

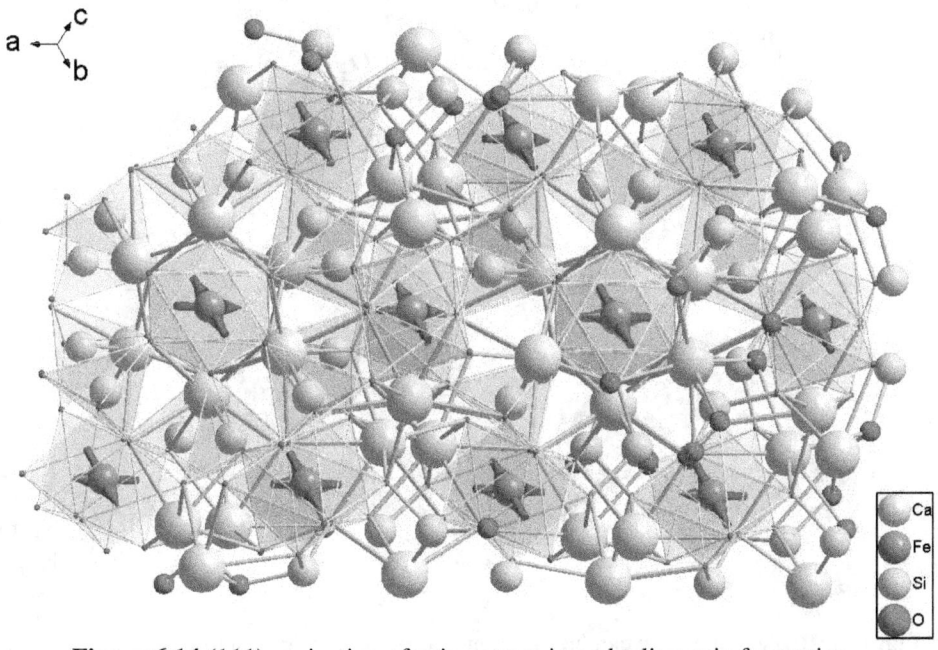

Figure 6.14 (111) projection of spin pattern in andradite: spin frustration.

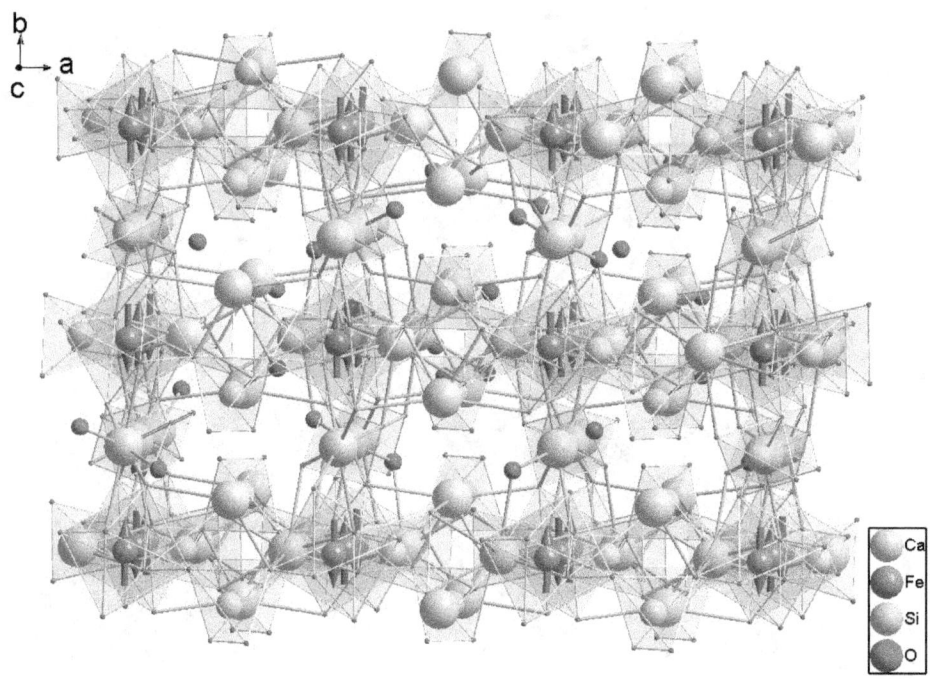

Figure 6.15 3D sublattice formed by the Fe octahedra connected via one Si tetrahedron and one Ca dodecahedron.

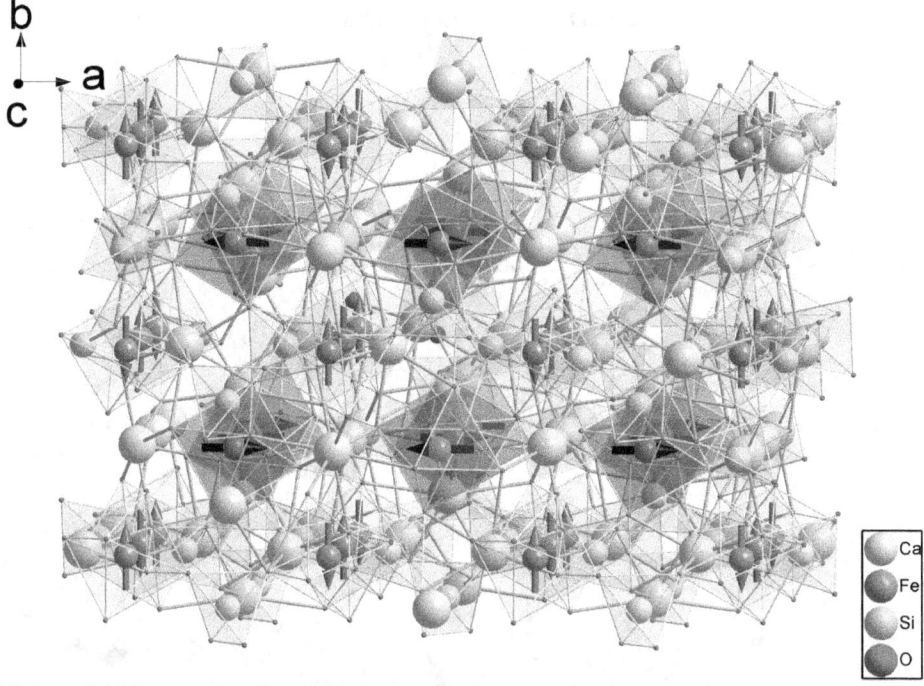

Figure 6.16 Two mutually canted interpenetrating magnetic sublattices connected via three Ca dodecahedra and two Si with one Ca polyhedra.

Based on the calculated spin-coupling constants and the consideration of the spin canting, it is possible to propose the spin frustrated pattern of andradite (Fig. 6.13). Within the frame of this pattern, spins are coupled antiparallel along each of the crystallographic axes and are collinear within the (100), (010) and (001) planes. The spins are canted along each of the space diagonals (111), (11-1), (1-11), (-111) and in every (111), (110), (101) and (011) planes (Fig. 6.14). Such structure can be considered as the magnetic structure of two mutually canted antiferromagnetic sublattices (Fig. 6.15). Within each antiferromagnetic sublattice spins are oriented antiparallel to next nearest neighbours and each pair of iron octahedra separated by 6.026 Å are connected via one Si and one Ca polyhedra (Fig. 6.15). The canting sublattices (Fig. 6.16) are connected via three Ca dodecahedra (Fig.6.11c) and two Si with one Ca polyhedra (Fig. 6.11b). Though, there are no experimental spin patterns in andradite, the calculated spin structure is in agreement with previous experimental investigations on andradite (Section 6.1).

Finally, each antiferromagnetic sublattice with collinear orientation of spins produces its own hyperfine pattern in the Mössbauer spectrum. The two sublattices are equivalent and therefore the Mössbauer parameters are almost identical, as it is. Therefore, though andradite structure has only one crystallographic Fe^{3+} site, the derived spin pattern of two canted antiferromagnetic sublattices explains two sextets in Mössbauer spectra below the Néel temperature.

6.4 Summary

The electronic and magnetic structure of andradite has been characterized by electronic structure calculations in the local spin density approximation. Whereas it is not possible to compare the calculated d-d energies with the experimental data due to unoccupancy of the $3d\downarrow$ MO, the calculated magnetic moments, isomer shifts and quadrupole splittings are in quantitative agreement with the corresponding experimental values, allowing the conclusion that the structure of andradite is correctly described, at least qualitatively. The t_{2g}-like orbitals are split in two levels and the e_g-like orbitals are degenerated showing that the Fe^{3+} polyhedron is the trigonally compressed octahedron (Section 3.1) Moreover, the consideration of the symmetry of the octahedron reveals the D_{3d} symmetry at 100 K. The positively charged clusters result the overestimation in the splitting of $3d$ orbitals and the isomer shift, whereas the negatively charged clusters result the overestimation in the quadrupole splitting. The substantial deviations of the calculated effective charges for Fe^{3+} and O ions from the formal oxidation states of +2 and −2, respectively, indicate considerable covalent character of the iron-oxygen bonds within the structure of andradite which is larger for the positively charged clusters.

The derived spin structure contains two mutually canted equivalent sublattices. Spins within every sublattice are antiferromagnetically coupled with next nearest neighbours. The

Fe octahedra within sublattice are separated by 6.026 Å and connected via one Si and one Ca polyhedra. In turn, the sublattices are connected via three Ca dodecahedra and two Si with one Ca polyhedra. The derived spin pattern of two canted antiferromagnetic sublattices is in agreement with the experiments and explains two sextets in Mössbauer spectra of andradite below the Néel temperature.

Chapter 7

SUMMARY AND CONCLUSIONS

The electronic structure calculations of orthoferrosilite $Fe^{2+}_2Si_2O_6$, almandine $Fe^{2+}_3Al_2(SiO_4)_3$ and andradite $Ca_3Fe^{3+}_2(SiO_4)_3$ have shown that a key issue of such cluster calculations is size convergence. In general, clusters which are size-converged with regard to optical transitions, hyperfine parameters and magnetic coupling constants data usually contain between 100 and 150 atoms, and represent the correct surroundings of the central Fe ion within a sphere with radius of about 5-6 Å. These clusters contain at least all coordination polyhedra of the cations bonded to the oxygen atoms of the first coordination sphere of the central transition metal ion.

The calculated effective charges for central Fe cations, as well as those for oxygen atoms of the first coordination sphere, exhibit substantial deviations from the formal oxidation states indicating considerable covalent character of the iron–oxygen bonds in the minerals. In addition, it was shown for the investigated Fe-bearing silicates that about 90% of the total overlap populations arise from the orbital interactions of the (formally empty) iron $4s$, $4p$ orbitals with the valence orbitals of the oxygens, whereas the competely filled $3d\uparrow$ orbitals yield a small antibonding contribution. Accordingly, the valence shell occupations of the iron cations contain considerable amounts from the $4s$ and $4p$ orbitals. These conclusions concerning considerable covalent character of the iron–oxygen bonds, total overlap populations can be generalized for Fe-bearing silicates.

In the case of orthoferrosilite (Chapter 4), the calculated quadrupole splitting, isomer shifts and the calculated d–d excitation energies for the size converged clusters around Fe(M1) and Fe(M2) are in quantitative agreement with the experimental values within the error margins expected due to the general theoretical approximations. This allows the conclusion that the electronic structure of orthoferrosilite is correctly being described, at least qualitatively. Next, with regard to the electronic structure, the four empty spin-down molecular orbitals with $3d$ character are mixtures of all five atomic $3d$-orbitals so that the assumption of a tetragonally compressed M1 octahedron with approximate symmetry D_{4h} is not justified. Finally, the calculated temperature dependence of the quadrupole splitting and asymmetry parameter, which arises from thermal population of the low lying excited molecular orbitals, is in quantitative agreement with the experimental data.

Analogous calculations on a synthetic Mg-rich orthopyroxene, $(Mg_{0.75}Fe_{0.25})_2Si_2O_6$ and a natural mantle orthopyroxene with small iron content (0.023 apfu in M1 and 0.127 apfu in M2) confirm previous conclusions regarding the calculations of spectroscopic data of minerals with low iron content that are of limited reliability due to local distortions of the environment of iron that are not properly accounted for in experimental structure determinations by diffraction methods.

111

The magnetic interactions between the iron spins within the ribbons and between neighboring ribbons in the (b, c)-plane are ferromagnetic in accordance with the experimental results. The coupling between different (b, c)-planes is antiferromagnetic via Si^B tetrahedra but ferromagnetic via the Si^A tetrahedra. Such a spin structure that may be denoted as "double-plane antiferromagnetic" resembles the pattern obtained recently for $LiFeGe_2O_6$, it is at variance with the magnetic structure derived from the neutron diffraction data. It is argued that this discrepancy may be attributed to differences in the $Fe(M2)$–O_3 distances at room temperature, that had to be used in the calculations, and the geometry below the Néel temperature. Hence, as soon as low temperature structural data are available, this problem may be resolved.

In the case of almandine (Chapter 5), the d-d excitation energies are somewhat smaller than the respective experimental values, but still are in reasonable agreement. The calculated magnetic moment, isomer shift and quadrupole splitting are in quantitative agreement with the corresponding experimental values, allowing the conclusion that the structure of almandine is correctly described, at least qualitatively. It was shown, that the splitting scheme of the spin-down $3d$ orbitals of Fe^{2+} in the crystal field of the almandine crystal structure differs from the splitting scheme in the field of the triangular dodecahedron with D_2 symmetry, and therefore cannot be described as the latter. Furthermore, the origin of the antiferromagnetism within the structure of almandine was revealed. The Fe edge-shared dodecahedra form two identical interpenetrating 3D sublattices of circles of 10 edge-shared dodecahedra. These two sublattices are connected via Al octahedra and Si tetrahedra. The magnetic interaction between the iron spins via oxygen bridges within each sublattice is ferromagnetic, whereas the coupling between two magnetic sublattices is weakly antiferromagnetic via Al octahedra and Si tetrahedra. This antiferromagnetic interaction of two identical magnetic sublattices is in qualitative agreement with experiment, and explains and solves the controversy in the interpretation of the Mössbauer spectra of almandine below the Néel temperature.

In Chapter 6, the electronic structure of andradite has been characterized. Whereas it is not possible to compare the calculated d-d energies with the experimental data due to unoccupancy of the $3d\downarrow$ MO, the calculated magnetic moment, isomer shift and quadrupole splitting are in quantitative agreement with the corresponding experimental values, allowing the conclusion that the structure of andradite is correctly described, at least qualitatively. The splitting pattern of the $3d\downarrow$ MO shows that the Fe^{3+} polyhedron corresponds to a trigonally compressed octahedron with D_{3d} symmetry at 100 K. In the next step, the magnetic structure of andradite was derived. The spin structure of andradite contains two mutually canted equivalent sublattices. The Fe octahedra within each sublattice are separated by 6.026 Å and connected via one Si and one Ca polyhedra. Each spin within every sublattice is antiferromagnetically coupled with next nearest spins. These two magnetic sublattices are connected via three Ca dodecahedra and two Si with one Ca polyhedra. The derived spin pattern of two canted antiferromagnetic sublattices is in agreement with the experiments and explains two sextets in Mössbauer spectra of andradite below the Néel temperature.

The calculating method can be successfully applied to other silicates and minerals in order to characterize and understand their electronic structure, magnetic structures and other properties, which is necessary for a deep understanding of the physics and chemistry of our planet and Universe.

From the academical point of view, we propose to calculate and create a database for splitting patterns, AO contributions and other characteristics of the $3d$ orbitals of Fe^{2+} ion at variable distortions (tetragonal, trigonal, angular etc.) of the distorted octahedral site since variable degrees of the distortions were calculated and analyzed. It was shown, that the point charge model for description of the crystal-field splitting is inefficient and the model based on the overlap integrals is very appropriate for the description of the crystal-field splitting.

Moreover, it would be great to create a database, which contained converged model clusters describing electronic structure and magnetic interactions for silicates and other minerals, like American Mineralogical Crystal Structure Database for X-ray diffraction data.

APPENDIX 1

Atomic input parameters.

Atom	Orbital	$\varepsilon_{nl\,0}$	$\varepsilon_{nl\,1}$	$\varepsilon_{nl\,2}$	$\zeta_{nl\,0}$	$\zeta_{nl\,1}$	η_0	η_1
Fe^{2+}							2.35	0.10
	4s	0.48	0.45	0.07	1.35	0.20		
	4p	0.20	0.36	0.07	2.60	0.50		
	3d	0.79	0.48	0.13	2.78	0.10		
Fe^{3+}							2.35	0.10
	4s	0.48	0.45	0.07	1.65	0.20		
	4p	0.15	0.36	0.07	2.80	0.60		
	3d	0.79	0.48	0.13	3.05	0.10		
O							2.55	1.30
	2s	2.30	1.58	0.26	2.30	0.10		
	2p	1.00	1.37	0.26	2.07	0.30		
F							2.90	3.30
	2s	2.82	1.88	0.26	2.65	0.10		
	2p	1.25	1.21	0.26	2.11	0.30		
Si							1.60	0.40
	3s	1.05	0.81	0.12	1.79	0.05		
	3p	0.49	0.64	0.12	1.25	0.20		
Al							1.55	0.40
	3s	0.78	0.60	0.12	1.50	0.05		
	3p	0.38	0.40	0.12	1.12	0.20		
Ca							1.35	0.40
	4s	0.30	0.20	0.06	1.07	0.00		
	4p	0.05	0.10	0.06	1.10	0.30		
Mg							1.45	0.40
	3s	0.56	0.45	0.12	1.10	0.05		
	3p	0.24	0.20	0.12	1.03	0.40		

APPENDIX 2

Data Input for the SCC-Xα programs.

The input file comprises the following sections:

(1) Program control (4 lines).

(2) Specification of atoms and coordinates.

(3) Atomic orbital (AO) data.

(4) Specification of AO-occupation numbers.

(5) Occupation numbers of partially occupied MO`s (optional).

(6) Symmetry coordinates (optional).

(7) Supplementary input for calculating spectroscopic data (optional).

Section 1: Program control.

First line:

column 1: a number between 1 und 4 specifies the type of calculation:

$1 \triangleq$ spin-restricted version without symmetry;

$2 \triangleq$ spin-polarized version without symmetry;

$3 \triangleq$ spin-restricted version with symmetry;

$4 \triangleq$ spin-polarized version with symmetry.

columns 2-73: Title for job identification (optional).

Second line: FORMAT 24I4, physical specification of the problem.

columns	notation	meaning
1 – 3	NA	number of atoms
4 – 6	NDA	number of different atoms
7 – 9	NORB	number of (spatial) atomic valence orbitals
10 – 12	NCHM	charge of the system
13 – 15	NE1UP	number of electrons in fully occupied (spin-up) molecular orbitals
16 – 18	NEVUP	number of partially occupied (spin-up) molecular orbitals
19 – 21	NE1DN	number of electrons in fully occupied spin-down molecular orbitals
22 – 24	NEVDN	number of partially occupied spin-down molecular orbitals

Third line: FORMAT 6I3,3F6.2,F7.4, numerical specification of the problem.

columns	notation	default	meaning
1 – 3	LLMAX	-	upper limit of l-summation in the three-center integrals
4 – 6	NSTEP	6	mesh for (radial) numerical integration
7 – 9	NIL	4	number of large iterations
10 – 12	NIS	9	number of small iterations
13 – 15	NUNIT	-	length unit for geometry input
16 – 18	NCAV	-	reduction of cluster charge for negatively charged systems
19 – 24	RNM	13.5	largest distance for computing two-center integrals
25 – 30	RNM3	8.1	largest distance for computing three-center integrals
31 – 36	DPF	0.15	damping factor for SCC iterations
37 – 43	CMIN	0.0001	convergency threshold

Fourth line: FORMAT 24I3, options for output and calculation of spectroscopic data, stored on array NPRI

columns	values	meaning
1 – 3		symmetry specification (> 0: symmetry basis used)
	0	no symmetry used
	1	symmetry basis used without additional output
	2	output list of the symmetry adapted basis
	3	as 2 + output of blockdiagonalized Hamiltonian
4 – 6		calculation of binding energy
	0	no calculation
	1	output of binding energies (in Ry, eV and kcal/mole)
	2	as 1 + output of total energy contributions
7 – 9		output options for SCC matrices
	0	minimal output (recommended during SCC iteration)
	1	output of atomic orbital list and diagonal elements of H
	2	+ output of overlap matrix (at the end of SCC iteration)
	3	+ output of V (ignored if NIT > 1)
	4	+ output of H (ignored if NIT > 1)
10 – 12		output options for final charge distribution
	0	atomic charge matrix only (recommended during SCC iteration)
	1	+ output of (nl)-AO-contributions, (valencies + bond indices)
	2	+ output of bond-charge matrix
	3	+ output of eigenvectors
	4	+ output of bond-order matrix
13 – 15		options for population analysis
	0	Mulliken population analysis
	1	population analysis weighted by ionic radii
	2	population analysis by numerical integration
16 – 18		options for calculating moments
	0	no calculation
	1	dipole moment
	2	+ second moments
19 – 21		options for calculating inverse moments
	0	no calculation
	n	inverse moments up to order $-n$ (current state: $n \leq 3$)

116

22 – 24		options for ESCA chemical shifts (additional input required)
	1	output of binding energy shifts (in eV)
	2	as 1 + output of contributions to the chemical shift
25 – 27		option for Mössbauer data (additional input required)
	0	no calculation
	1	efg and $\rho(0)$, without output of S_{cv}
	2	efg and $\rho(0)$, with output of S_{cv}
	3	efg only
	4	$\rho(0)$, with output of S_{cv}
	5	$\rho(0)$, without output of S_{cv}

Section 2: Specification of atoms and coordinates.

Each atom of the molecule has to be specified on a separate line; FORMAT 2I3,A4,3F10.6,3F7.3.

columns	notation	meaning	
1 – 3	NAT	number of the atom in the cluster	
4 – 6	NORD	atomic number (= nuclear charge)	
7 – 10	NSYM	"chemical symbol"	
11 – 20		x–	
21 – 30		y–	coordinates (in a.u. or Å)
31 – 40		z–	
41 – 47	ALFA	η_0	potential parameter coefficients
48 – 54	DALFA	η_0	with $\eta = \eta_0 + \eta_1 Q$
55 – 61	TDS	kinetic energy destabilization	

Section 3: Atomic orbital (AO) data.

The atoms with different chemical symbols are specified by their (nl)-orbitals, usually only valence orbitals: FORMAT A4,I1,2I2,2X,6F7.4,I3

columns	notation	meaning	
1 – 4	NSYM	chemical symbol from the atom list	
5	NDO	number of different (nl)-orbitals (≤ 8)	
6 – 7	NF	main quantum number n	
8 – 9	LF	angular moment quantum number l	
12 – 18	EAT	ε_0	atomic orbital energy
19 – 25	FAT	ε_1	coefficients (in Ry):
26 – 32	GAT	ε_2	$\varepsilon = \varepsilon_0 + \varepsilon_1 Q + \varepsilon_2 Q^2$
33 – 39	ZW (1)	ζ_0	atomic orbital exponents:
40 – 46	ZW (2)	ζ_1	$\zeta = \zeta_0 + \zeta_1 Q$
47 – 53		Intra-atomic exchange splitting (Stoner) parameter	
54 – 56		atomic orbital occupation number in the ground state	

Section 4: Specification of AO-occupation numbers and spin densities.

The occupation numbers (and spin densities for spin-polarized calculations) of the atomic (nl)-orbitals supply the starting values for the iteration process: FORMAT 10F8.6; continuing cards are possible according to the total number of atomic (nl)-orbitals; first set: atomic (nl) occupation numbers; second set: atomic (nl) spin densities.

In principle, the ground state atomic orbital occupation numbers (given with the atomic orbital cards) may be used as starting point, but for complex systems, especially with transition metals, such a choice may converge slowly or not at all.

Section 5: Occupation numbers of partially occupied MO`s (optional).

If NEVUP and/or NEVDN $\neq$ 0, the program requests NEVUP and/or NEVDN MO occupation numbers x_k to be specified ($0 \leq x_k \leq 2$ spin-restricted, $0 \leq x_k \leq 1$ spin-polarized): FORMAT 10F8.6.

If neither symmetry is being used nor any other input for spectroscopic data is required, the job is terminated either with "0" in third column (no more jobs) or with "1" if another data set is to be calculated.

Section 6: Symmetry coordinates (optional).

If NPRI (1) $\neq$ 0, the program requires the specification of the symmetry group of the system and the list of the symmetry coordinates (SYCO`s). The (unnormalized) SYCO`s are read in an INTEGER-Format. If the system has 3-, 5- or higher-fold symmetry, not all of the SYCO`s can be represented as integers. In that case, the input-integers have to be converted to real characters.

First line: FORMAT A4,2I4
column 1 – 4: group symbol (MBG, e.g. C3V for C_{3v}).
column 5 – 8: number of actually occurring irreducible representations (NREP).
column 9 – 12: number of different integers to be converted (NUM).

Second line: FORMAT 2F8.5 (optional)
If NUM $\neq$ 0, NUM lines are required with the pairs of numbers to be converted into each other, e.g.:
_3.00000_1.73205
- 3.00000-1.73205 etc.
No cards are required if NUM = 0

Subsequent lines: for each irreducible representation the SYCO`s are listed beginning with a line denoting the irreducible representation, e.g. TIGX for the first row of $t_{1g}$ in $O_h$, together with the number of SYCO`s transforming according to this representation, i.e.:

<u>One line:</u> FORMAT A4,I4

column 1 – 4: symbol for the irreducible representation.

column 5 – 8: number of SYCO`s belonging to this representation.

<u>Next cards:</u> FORMAT 36I2 (+ continuing cards if NORB < 36), list of SYCO`s.

Each SYCO`s requires $int_+(NORB/36)$ lines, and each SYCO starts on a new line.

Section 7: Supplementary input for calculating spectroscopic data (optional).

The calculation of some of the spectroscopic data requires additional input, especially the hyperfine data required for the interpretation of Mössbauer spectra.

The example of input file is in the next page.

File   Edit   Search   Preferences   Shell   Macro   Windows

```
2_Fe(II)O6-Si6O2F14-Mg5F6(nt):SC76M1-40_____RT_____corrected_____
  40   7 165   0 117   0 113   0
   6   8   1   1   1 13.50  8.25  0.15 .00002
   0   0   0   1   1   0   0   3
     1 26FE      0.000000  0.000000  0.000000   2.35  0.10  0.23      0.00000
    34  8O      -1.114842  0.056737 -1.758974   2.55  1.30  0.00      2.08329  O1A(Si,Fe)
    49  8O      -1.177463 -1.432054  0.961440   2.55  1.30  0.00      2.08844  O2A(Si,Fe,Mg)
   457  8O       1.058482 -1.575487 -0.949919   2.55  1.30  0.00      2.12247  O2B(Si,Fe)
    45  8O       1.165307  0.084879  1.773113   2.55  1.30  0.00      2.12346  O1B(Si,Fe,Mg)
    43  8O       1.165307  1.655373 -0.845187   2.55  1.30  0.00      2.19375  O1B(Si,Fe)
    40  8O      -1.114842  1.683515  0.859326   2.55  1.30  0.00      2.19443  O1A(Si,Fe)
   425 14SI      1.793361 -2.901964 -0.447729   1.60  0.40  0.35      3.44064
   417 14SI     -1.905158 -2.863655  0.914834   1.60  0.40  0.35      3.55908
    18 14SI     -2.726232  0.064908 -1.703466   1.60  0.40  0.35      3.21533
    24 14SI     -2.726232  1.675345  0.914834   1.60  0.40  0.35      3.32807
   107 14SI      2.784249  1.637036 -0.857231   1.60  0.40  0.35      3.34167
   109 14SI      2.784249  0.103217  1.761069   1.60  0.40  0.35      3.29607
     7 12MG      0.000000  1.740253  2.618300   1.45  0.40  0.55      3.14388  2*O,3*FA
   409 12MG      0.035547 -1.529461  2.576931   1.45  0.40  0.55      2.99685  2*O,2*FS,FA
    15 12MG      0.035547  3.269714 -0.041369   1.45  0.40  0.55      3.27017  2*O,2*FA
  1449 12MG      0.035547 -1.529461 -2.659669   1.45  0.40  0.55      3.06828  2*O,2*FA
   487 12MG      0.000000  1.740253 -2.618300   1.45  0.40  0.55      3.14388  2*O,3*FA
    72  8OA     -3.282640  0.745758 -0.305294   2.55  1.30  0.00      3.38010  O3A(Si,Si)
   155  8OA      3.253908  0.441645  0.203704   2.55  1.30  0.00      3.29006  O3B(Si,Si)
   139  9FS      3.519127  2.963513 -0.355041   2.90  3.00  0.00      4.61440  O2B(Si)
   141  9FS      3.519127 -1.223260  2.263259   2.90  3.00  0.00      4.35924  O2B(Si)
    56  9FS     -3.453928  3.106946  0.961440   2.90  3.00  0.00      4.74416  O2A(Si)
   450  9FS     -3.453928 -1.366693 -1.656860   2.90  3.00  0.00      4.06727  O2A(Si)
   433  9FS     -3.516549 -2.855485  0.859326   2.90  3.00  0.00      4.61068  O1A(Si)
   761  9FS      3.412303 -2.883627 -0.459773   2.90  3.00  0.00      4.49116  O1B(Si)
   465  9FS     -1.348750 -3.793242 -0.305294   2.90  3.00  0.00      4.03745  O3A(Si)
   546  9FS     -3.282640  0.994495 -2.923594   2.90  3.00  0.00      4.50690  O3A(Si)
   479  9FS      1.323702 -3.240392  1.109636   2.90  3.00  0.00      3.67200  O3B(Si,Mg)
   877  9FS      3.253908  1.298608 -2.414596   2.90  3.00  0.00      4.25495  O3B(Si)
    66  9FS     -3.282640  0.994495  2.313006   2.90  3.00  0.00      4.13700  O3A(Si)
   471  9FS     -1.348750 -3.544505  2.313006   2.90  3.00  0.00      4.44214  O3A(Si,Mg)
   157  9FS      3.253908  1.298608  2.822004   2.90  3.00  0.00      4.49867  O3B(Si)
   473  9FS      1.323702 -4.097355 -1.508664   2.90  3.00  0.00      4.56252  O3B(Si)
  1257  9FA      1.058482 -1.575487  4.286681   2.90  3.30  0.00      4.68809  O2B(Mg)
    55  9FA     -1.177463  3.172307 -1.656860   2.90  3.30  0.00      3.76764  O2A(Mg,Mg)
   529  9FA     -1.177463 -1.432055 -4.275160   2.90  3.30  0.00      4.65985  O2A(Mg)
   295  9FA     -1.177463  3.172307  3.579740   2.90  3.30  0.00      4.92590  O2A(Mg)
    63  9FA      1.058482  3.315740  1.668301   2.90  3.30  0.00      3.85979  O2B(Mg,Mg)
   543  9FA      1.058482  3.315739 -3.568219   2.90  3.30  0.00      4.98465  O2B(Mg)
FE   3  4  0   0.48   0.45   0.07   1.35   0.20   0.06   2
FE      4  1   0.20   0.36   0.07   2.60   0.50   0.06   0
FE   3     2   0.79   0.48   0.13   2.78   0.10   0.13   6
O    2  2  0   2.30   1.58   0.26   2.30   0.10   0.30   2
O       2  1   1.00   1.37   0.26   2.04   0.30   0.30   4
SI   2  3  0   1.05   0.81   0.12   1.79   0.05   0.08   2
SI      3  1   0.49   0.64   0.12   1.25   0.20   0.10   2
MG   2  3  0   0.56   0.45   0.12   1.10   0.05   0.15   2
MG      3  1   0.24   0.20   0.12   1.03   0.40   0.15   0
OA   2  2  0   2.30   1.58   0.26   2.30   0.10   0.30   2
OA      2  1   1.00   1.37   0.26   2.07   0.30   0.30   4
FS   2  2  0   2.82   1.88   0.26   2.65   0.10   0.30   2
FS      2  1   1.25   1.41   0.26   2.11   0.30   0.30   5
FA   2  2  0   2.82   1.88   0.26   2.65   0.10   0.30   2
FA      2  1   1.25   1.21   0.26   2.11   0.30   0.30   5
0.5388360.3841986.4420331.4361314.8892131.4080084.9643271.3989424.9458291.436862
4.9034211.4262224.9255361.4198224.9307701.1113132.2000801.1098432.2190821.112554
2.2159331.1179982.2321581.1179892.2137081.1180082.2088000.4879640.9741920.469649
0.8815700.4311530.8452050.4876710.8599490.4902500.9742721.3310114.9879161.356083
4.9726551.6311615.5635251.6283795.5619211.6320755.5709271.6305635.5655051.639154
5.5494121.6487025.5434341.6508875.5530401.6502835.5587071.6703815.5318091.666506
5.5512571.6551205.5720351.6720355.5332601.6721255.5543431.6753915.5420771.674721
5.6998341.6805745.6144711.6888075.6794781.6907895.7143641.6768035.6163681.702649
5.709971
 0.01696 0.06122 3.52493 0.00184 0.06032 0.00402 0.07495 0.00313 0.05906 0.00209
 0.06135 0.00206 0.05297 0.00216 0.05731-0.00032-0.00028-0.00011 0.00010-0.00061
-0.00091-0.00057-0.00077-0.00045-0.00066-0.00054-0.00084-0.00140-0.00128-0.00153
-0.00116-0.00106-0.00026-0.00158-0.00041-0.00109-0.00110-0.00011 0.00178-0.00008
 0.00209 0.00017 0.00129 0.00009 0.00087 0.00022 0.00155 0.00005 0.00080 0.00016
 0.00145 0.00012 0.00110 0.00012 0.00075 0.00020 0.00184-0.00003 0.00064 0.00004
 0.00101 0.00001 0.00070 0.00016 0.00306 0.00016 0.00184 0.00023 0.00219 0.00005
 0.00111 0.00001 0.00178 0.00002 0.00136-0.00003 0.00104 0.00000 0.00180-0.00007
 0.00091
  1    MOESSBAUER  DATA  FOR  IRON
FE   1  0  8
-11.000  0.15    4.25    1.3    0.14 102.0    0.19   1.06   14.0    0.71
  8.71  -0.052  -0.604   0.     0.     0.      0.     0.     0.      0.
  0       FORTSETZUNGSKARTE
```

REFERENCES

Adetunji J. and Dronsfield A. (2002): The beginnings of Mössbauer spectroscopy. Education in Chemistry, 39 (4), p. 97–100.

American Mineralogist Crystal Structure Database,
http://rruff.geo.arizona.edu/AMS/amcsd.php

Amthauer G., Annersten H. and Hafner S.S. (1976): The Mössbauer spectrum of ^{57}Fe in silicate garnets. Zeit. Kristallographie, 143, p. 14–55.

Amthauer G., Grodzicki M., Lottermoser W. and Redhammer G. (2004): Mössbauer spectroscopy: Basic principles. EMU Notes in Min. 6, p. 345–367.

Anderson P. (1950): Antiferromagnetism. Theory of Superexchange Interaction. Phys. Rev. 79 (2), p. 350–356.

Anderson P. (1959): New Approach to the Theory of Superexchange Interactions. Phys. Rev. 115 (1), p. 1–13.

Anderson P. (1963): Theory of Magnetic Exchange Interactions: Exchange in Insulators and Semiconductors. Sol. State. Phys. 14, Academic Press, New York, p. 99–214.

Anderson P. (1967): Magnetism. Vol. I. Academic Press, New York, USA, p. 67.

Andrut M., Wildner M. and Rudowicz C. Optical absorption spectroscopy in geosciences. Part II: Quantitative aspects of crystal fields. EMU notes in mineralogy. Eötvös University Press, Budapest, 2004, p. 145–188.

Anovitz L., Essene E., Metz G., Bohlen S., Westrum Jr. E. and Hemingway B. (1993):: Heat capacity and phase equilibria of almandine, $Fe_3Al_2Si_3O_{12}$. Geochim. Cosmochim. Acta 57, p. 4191–4204.

Anslyn E., Dougherty D. Modern Physical Organic Chemistry. University Science Books, 2006.

Armbruster T. and Geiger C. (1993): Andradite crystal chemistry, dynamic X-site disorder and structural strain in silicate garnets. Eur. J. Mineral 5, p. 59–71.

Ashcroft N. and Merin N. Solid State Physics. Harcourt College Publishers, USA, 1976.

Bancroft G.M. and Burns R.G. (1967): Interpretation of the electronic spectra of iron in pyroxenes. Am. Mineral. 52, 1278–1287.

Barnett M. and Coulson C. (1951): The Evaluation of Integrals Occurring in the Theory of Molecular Structure. Parts I & II. Phil. Trans R. Soc. Lond. A 243, p. 221–249.

Basch H., Viste A. and Gray H. (1965): Valence Orbital Ionization Potential from atomic spectral data. Theor. Chim. Acta 3, p. 458–464.

Born M. and Oppenheimer J. (1927): Zur Quantentheorie der Molekeln. Ann. Physik 84, p. 457–484.

Bethe H. (1929): Termaufspaltung in Kristallen. Ann. Physik 395 (2), p.133–276.

Blaha P., Schwarz K. and Luitz J. (2000): Calculations of electric field gradient in solids: how theory can complement experiment. Hyperfine Int. 126, p. 389–395.

Burnham C.W., Ohashi Y., Hafner S. S., Virgo D. (1971): Cation distribution and atomic thermal vibrations in an iron-rich orthopyroxene. Am. Mineral. 56, p. 850–876.

Burns R. (1970): Crystal field spectra and evidence of cation ordering in olivine minerals. Amer. Mineral. 55, p. 1608–1632.

Burns R.G. (1981): Intervalence transitions in mixed-valence minerals of iron and titanium. Ann.Rev.Earth.Planet.Sci. 9, p. 345–383.

Burns R. Mineralogical applications of crystal field theory. 2nd Ed., Cambridge University Press, Cambridge, 1993.

Ceperley D. and Alder B. (1980): Ground State of the Electron Gas by a Stochastic Method. Phys. Rev. Lett. 45, p.566 – 569.

Chen C. Magnetism and metallurgy of soft magnetic materials. Dover Publications. Toronto, Canada, 1986.

Clark S. (1957): Absorption spectra of some silicates in the visible and near infrared. Am. Mineral. 42, p. 732–742.

Clementi E. and Roetti C. (1974): Roothan-Hartree-Fock atomic wavefunctions. At. Nucl. Data. Tables 14, p.177–478.

de Oliveira J.C.P., Kunrath J., and Vasquez A. (1987): The Mössbauer effect in natural almandine: $Fe_3Al_2(SiO_4)_3$. *Phys. Scripta* 36, p. 526–528.

de Oliveira J.C.P., da Costa M., Schreiner W. and Vasquez A. (1989): Magnetic properties of the natural pyrope-almandine garnets. *J. Magn. Magn. Mater.* 79, p. 1–7.

Deer W., Howie R. and Zussman J. Orthosilicates. Geological Society, Bath, UK, 1997.

Deer W.A., Howie R.A., Zussman J. Rock-forming minerals, vol. 2A. Longman, London, 2001.

Diamond 3, Crystal Structure Visualization Software,

http://www.crystalimpact.com/diamond/Default.htm

Diego Gatta G., Rinaldi R., Knight K.S., Molin G., Artioli G. (2007): High temperature structural and thermoelastic behaviour of mantle orthopyroxene: an in situ neutron powder diffraction study. Phys. Chem. Minerals 34, p. 185–200.

Dufek P., Blaha P. and Schwarz K. (1995): Determination of the nuclear quadrupole moment of ^{57}Fe. Phys.Rev.Let. 75, p. 3545–3548.

Edmonds A.R. Drehimpulse in der Quantenmechanik. Hochschultaschenbücher-Verlag, Manheim, Germany, 1964.

Eméléus H., Sharpe A. Advances in inorganic chemistry and radiochemistry. Vol.20. Academic Press, New York, 1977.

Fermi, E. (1927): Un Metodo Statistico per la Determinazione di alcune Priorieta dell'Atome. Rend. Accad. Naz. Lincei, 6, p. 602–607.

Gaft M., Reisfeld R., Panczer G. Luminescence spectroscopy of minerals and materials, Springer-Verlag Berlin Heidelberg, 2005.

Geiger C., Armbruster Th., Lager G., Jiang K., Lottermoser W., and Amthauer G. (1992): A Combined Temperature Dependent ^{57}Fe Mössbauer and Single crystal X-ray Diffreaction Study of Synthetic Almandine: Evidence for the Gol'danskii-Karyagin Effect. Phys. Chem. Minerals 19, p. 121–126.

Geiger C. and Rossman G. (1994): Crystal field stabilization energies of almandine–pyrope and almandine–spessartine garnets determined by FTIR near-infrared measurements. Phys. Chem. Minerals 21, p. 516–525.

Geiger C. and Feenstra A. (1997): Molar volumes of mixing of almandine-pyrope and almandine–spessartine garnets and the crystal chemistry and thermodynamic-mixing properties of the aluminosilicate garnets. Am. Mineral. 82, p. 571–581.

Geiger C.A., Grodzicki M., Amthauer G. (2003): The crystal chemistry and FeII–site properties of aluminosilicate garnet solid solutions as revealed by Mössbauer spectroscopy and electronic structure calculations. Phys. Chem. Minerals 30, p. 280–292.

Geller, S. (1967): Crystal chemistry of the garnets. Zeit. Kristallographie, 125, p. 1 –47.

Gerloch M. and Slade R. Ligand-field parameters. Cambridge University Press. Cambridge, 1973.

Ginsberg A. (1971): Magnetic exchange in transitional metal complexes VI. Aspects of exchange coupling in magnetic cluster complexes. Inorg. Chim. Acta Rev. 5, p. 45–68.

Goldman D.S. and Rossman G. R. (1977): The spectra of iron in pyroxene revisited: the splitting of the ground state. Am. Mineral. 62, p.151–157.

Goldman D.S. and Rossman G. R. (1979): Determination of quantitative cation distribution in orthopyroxenes from electronic absorption spectra. Phys. Chem. Minerals 4, p. 43–53.

Goodenough J. (1958): An Interpretation of the Magnetic Properties of the Perovskite-Type Mixed Crystals. J. Phys. Chem. Solids 6, p. 287–297.

Grodzicki M. (1980): A self-consistent charge-Xα method. I. Theory. J. Phys. B13, p. 2683–2691.

Grodzicki M. (1980): A self-consistent charge-Xα method. II. Application to small molecules. J. Phys. B13, p. 2693–2700.

Grodzicki M., Lauer S., Trautwein A., Vera A. (1981): Application of Molecular Orbital Calculations to Mössbauer and NMR Spectroscopy of Halogen Containing Compounds. ACS Adv. Chem. Series 194, p. 3–37.

Grodzicki M., Walther H., Elbel S. (1984): The Electronic Structures of the Group V Series ER_3 (E = N$\rightarrow$Sb; R = H, Hal). An Intercomparison of Photoelectron Spectra and SCC-Xα Calculations. Zeit. für Naturforschung 39b, p. 1319-1330.

Grodzicki M. Theorie und Anwendungen der Self-Consistent-Charge-Xα Methode. Habilitation Thesis, Hamburg, 1985.

Grodzicki M., Männing V., Trautwein A., Friedt J. (1987): Calibration of isomer shifts and quadrupole coupling constants for ^{119}Sn, ^{127}I and ^{129}I as derived from SCC-Xα calculations and Mössbauer measurements. J. Phys. B 20, p. 5595–5625.

Grodzicki M. Dichtefunktionaltheorie. Essay in: Lexikon der Physik, 2, p. 18–22, Spectrum Verlag, Heidelberg, 1999.

Grodzicki M. and Amthauer G. (2000): Electronic and magnetic structure of vivianite: cluster molecular orbital calculations. Phys. Chem. Minerals 27, p. 694–702.

Grodzicki M., Heuss–Assbichler S. and Amthauer G. (2001): Mössbauer investigations and molecular orbital calculations on epidote. Phys. Chem. Minerals 28, p. 675–681.

Grodzicki M., Redhammer G.J., Amthauer G., Schünemann V., Trautwein A.X.,Velickov B., and Schmid–Beurmann P. (2003): Electronic structure of Fe-bearing lazulites. Am. Mineral. 88, p. 481–488.

Grodzicki M. Lectures: Introduction to SCC–Xα method, 2007.

Grodzicki M. VO Physik III: Struktur der Materie, 2007.

Grodzicki M., Redhammer G., Reissner M., Steiner W., Amthauer G. (2009): Electronic and magnetic structureof pyroxenes I. Hedenbergite. Phys. Chem. Minerals 36, in press.

Grodzicki M. and Lebernegg S. Computation and Interpretation of Mössbauer Parameters of Fe–bearing Compounds. In Book: Mössbauer Spectroscopy and Transition Metal Chemistry: Fundamentals and Application. Springer, Berlin, 2010.

Gütlich P., Link R. and Trautwein A. Mössbauer spectroscopy and transition metal chemistry. Springer-Verlag, Berlin, 1978.

Hahn T. vol. A. Space group symmetry. International tables for crystallography. 2nd Ed., D. Reidel Publishing Company, Dodrecht, 1987.

Heisenberg W. (1926): Über die Spektra von Atomsystemen mit zwei Elektronen. Z. Phys. 39, p. 499–518.

Heisenberg W. (1928): Zur Theorie des Ferromagnetismus. Z. Phys. 49, p. 619–636.

Heitler W. and London F. (1927): Wechselwirkung neutraler Atome und homöopolare Bindung nach der Quantenmechanik. Zeit. Physik 44, p. 455–472.

Hinchliffe A. Molecular modeling for beginners. Wiley, GB, 2003.

Hiroi T., Takeda H. (1992): Crystal-field theory calculations for Fe^{2+} ions in bronzite, augite and olivine. Phys. Chem. Minerals. 19, p. 229–235.

Huggins F.E. (1975): The $3d$ levels of ferrous ions in silicate garnets. Am. Mineral. 60, p. 316–319.

Hohenberg P. and Kohn W. Inhomogeneous Electron Gas. Phys.Rev. 136, 1964, p.864–871.

Hugh-Jones D.A., Chopelas A., Angel R.J. (1997): Tetrahedral compression in $(Mg,Fe)SiO_3$ orthopyroxenes. Phys. Chem. Minerals 24, p. 301–310.

Jahn H. and Teller E. (1937): Stability of Polyatomic Molecules in Degenerate Electronic States. I. Orbital Degeneracy. Proc. R. Soc. Lond. A. 161 (905), p.220–235.

Jensen F. Introduction to computational chemistry. Wiley, GB, 1999.

Jibamitra G. and Subrata G. (1979): Aluminous Orthopyroxene: Order-Disorder, Thermodinamic Properties, and Petrologic Implications. Contrib. Mineral. Petrol. 69, p. 375–385.

Jodlauk S., Becker P., Mydosh J.A., Khomskii D.I., Lorenz T., Streltsov S.V., Hezel D.C., Bohaty L. (2007): Pyroxenes: a new class of multiferroics. J. Phys. Cond Matter 19, p.432201 (9 pages).

Kohn W. and Sham L. (1965): Self–Consistent Equations Including Exchange and Correlation Effects. Phys. Rev.140 (4A), p. 1133–1138.

Kohn W. (1995): Overview of Density Functional Theory. In book Density functional Theory. NATO ASI Series. Series B: Physics 337, p. 3–10.

Kohn W. (1999): Electronic Structure of Matter – Wave Functions and Density Functionals. Nobel Lecture, Rev. Modern Physics 71 (5), p. 213–237.

Kanamori, J. (1959): Superexchange Interaction and Symmetry Properties of Electron Orbitals. J. Phys. Chem. Solids 10, p. 87–98.

Keutel H., Käpplinger I., Jäger E.G., Grodzicki M., Schünemann V., Trautwein A.X. (1999): Structural, magnetic and electronic properties of a pentacoordinated intermediate-spin (S = 3/2) iron(III) complex with a macrocyclic $[N_4]^{2-}$ ligand. Inorg. Chem. 38, p. 2320–2327.

Kramers H. (1934): L'interaction Entre les Atomes Magnétogènes dans un Cristal paramagnétique. Physica 1, p. 182–192.

Kurian R., Filatov M. Calibration of ^{119}Sn isomer shift using ab initio wavefunction methods. ICAME 09, July 2009, Vienna, contribution T02–P03.

Kühberger A., Fehr T., Huckenholz H.G. and Amthauer,G. (1989): Crystal chemistry of a natural schorlomite and Ti-andradites synthesized at different oxygen fugacities. Phys. Chem. Minerals, 16, p. 734–740.

Lewars E. Computational Chemistry: Introduction to the Theory and Applications of Molecular and Quantum Mechanics. Kluwer Academic Publishers, Dordrecht, 2003.

Lougear A., Grodzicki M., Bertoldi C., Trautwein A.X. Steiner K. and Amthauer G. (2000): Mössbauer and molecular orbital study of chlorites. Phys. Chem. Minerals 27, p. 258–269.

Luo J., Xue Z., Liu W., Wu J., Yang Z. (2006): Koopman´s theorem for large molecular systems within density functional theory. J. Phys. Chem. A 110 (43), p. 12005–9.

Langer K., Khomenko V.M. (1999): The influence of crystal field stabilization energy on Fe^{2+} partitioning in paragenetic minerals. Contrib. Mineral. Petrol. 137, p. 220–231.

Lauer S., Marathe V.R. and Trautwein A.X. (1979): Sternheimer shielding using various approximations. Phys. Rev. A 19, p. 1852–1861.

Lebernegg S., Amthauer G. and Grodzicki M. (2008): Single-centre MO theory of transition metal complexes. J. Phys. B: At. Mol. Opt. Phys. 41 (3), p. 035102–035109.

Lebernegg S., Grodzicki M. Anwendung und Grenzen der Kristallfeldtheorie: Ein molekülorbital-theoretischer Vergleich. VDM Verlag Dr. Müller, Saarbrücken, 2009.

Lebernegg S. (2010): Personal conversation.

Lever A.B.P. Inorganic electronic spectroscopy. 2nd Ed. Elsevier, Amsterdam, 1984.

Lin C., Zhang L., Hafner S.S. (1993): Local electronic states of Fe^{2+} ions in orthopyroxenes. Am. Mineral. 78, p. 8–15.

Lottermoser W., Steiner K., Scharfetter G., Jiang K., Grodzicki M., Redhammer G., Amthauer G., Treutmann W. (2002): The electric field gradient in synthetic fayalite α-Fe_2SiO_4 at moderate temperatures. Phys. Chem. Minerals. 29, p. 112–121.

Lougear A., Grodzicki M., Bertoldi C., Trautwein A.X. Steiner K. and Amthauer G. (2000): Mössbauer and molecular orbital study of chlorites. Phys. Chem. Minerals 27, p. 258–269.

Manning P. (1967): The optical absorption spectra of the garnets almandine-pyrope, pyrope, and spessartite and some structural interpretations of mineralogical significance. Canad. Mineral. 9, p. 237–251.

Manning P. and Townsend M. (1969): Effect of next-nearest neighbor interaction on oscillator strengths in garnets. J. Phys. C: Solid State Phys. 3, L14–15.

Manning P. and Harris D. (1970): Optical-absorption and electron-microprobe studies of some high-Ti andradites. Canad. Mineral. 10, p. 260–271.

Manning P. and Tricker M. (1977): A Mössbauer spectral study of ferrous and ferric ion distributions in grossular crystals: Evidence for local crystal disorder. Canad. Mineral., 15, p. 8l–86.

Manning V. and Grodzicki M. (1986): Theoretical interpretation of Mössbauer spectra of ^{119}Sn compounds. Theor. Chem. Acta 70, p. 189–202.

McCammon C. (2004): Mössbauer spectroscopy: Applications, EMU Notes in Min. 6, p. 369–398.

Mitra S. Fundamentals of Optical, Spectroscopic ond X-Ray Mineralogy. New Age International, New Delhi, 1996.

Mössbauer R. L. (1958): Kernresonanzfluoreszenz von Gammastrahlung in Ir^{191}. Z. Physik 151, p.124–143.

Mössbauer R. L. (1958): Kernresonanzabsorption von Gammastrahlung in Ir^{191}. Naturwissenschaften 22, p. 538–539.

Mössbauer R. Biography from *Nobel Lectures, Physics 1942-1962*, Elsevier Publishing Company, Amsterdam, 1964.
http://nobelprize.org/nobel_prizes/physics/laureates/1961/mossbauer-bio.html

Mulliken R.S. (1955): Electronic Population Analysis on LCAO–MO Molecular Wace Functions. J. Chem. Phys. 23 (10), p. 1833–2343.

Murad E. (1984): Magnetic ordering in andradite. Am. Mineral. 69, p. 722–724.

Murad E. and Wagner F. (1987): The Mössbauer Spectrum of Almandine. Phys. Chem. Minerals 14, p. 264–269.

Noodleman L. (1981): Valence bond description of antiferromagnetic coupling in transition metal dimers. J. Chem. Phys. 74, p. 5737–5743.

Novak G.A. and Gibbs G.V. (1971): The Crystal Chemistry of the Silicate Garnets. Am. Mineral. 56, p. 791–825.

Paulsen H., Ding X., Grodzicki M, Butzlaff C., Trautwein A., Hartung R., Wieghardt K (1994): Spectroscopic and theoretical studies on a three-iron cluster with linear arrangement. Chem. Phys. 184, p. 149–162.

Perdew J. and Zunger A. (1981): Self–interaction correction to density–functional approximations for many–electron systems. Phys. Rev. B 23, p. 5048 – 5079.

Perdew J. and Wang Y. (1992): Accurate and simple analytic representation of the electron–gas correlation energy. Phys. Rev. B 45, p. 13244–9.

Prandl W. (1971): Die magnetische Struktur und die Atomparameter des Almandins $Al_2Fe_3(SiO_4)_3$. Z. Kristallogr. 134, p. 344–349.

Ray S. and Das T. (1977): Nuclear quadrupole interaction in the Fe^{2+} ion including many-body effects. Phys.Rev.B 16, p. 4794–4804.

Redhammer G., Roth G., Treutmann W., Hoelzel M., Paulus W., Andre G., Pietzonka C., Amthauer G. (2009): The magnetic structure of clinopyroxene-type $LiFeGe_2O_6$ and revised data on multiferroic $LiFeSi_2O_6$. J. Sol. State. Chem. 182, p. 2374–2384.

Regnard J.R., Guillen R., Wiedenmann A., Fillion G., Hafner S., Langer K. (1986): Mössbauer and magnetic studies of orthorhombic $FeSiO_3$. Hyperfine Int. 28, p. 589–592.

Ruiz E., Alemany P., Alvarez S. and Cano J. (1997): Towards the Prediction of Magnetic Coupling in Molewcular Systems: Hydroxo- and Alkoxo-Bridged Cu(II) Binuclear Complexes. J. Am. Chem. Soc. 119, p. 1297–1303.

Runciman W.A., Sengupta D. And Marshall M. (1973): The polarized spectra of iron in silicates. I. Enstatite. Am. Mineral. 58, p. 444–450.

Seifert F. (1983): Mössbauer line broadening in aluminous orthopyroxenes: Evidence for next nearest neighbors interactions and short-range order. N. Jb. Minerals Abh. 148, p. 141–162.

Shenoy G.K., Kalvius G.M., Hafner S.S. (1969): Magnetic behavior of the $FeSiO_3$-$MgSiO_3$ orthopyroxene system from NGR in ^{57}Fe. J. Appl. Phys. 40, p. 1314–1316.

Sherrill D. The Born–Oppenheimer Approximation, School of Chemistry and Biochemistry, March 2005.

Slater J.C. (1951): A simplification of the Hartree–Fock method. Phys. Rev. 81, p. 385–390.

Slater J.C. Quantum theory of molecules and solids, Vol.4 McGraw-Hill, New York, 1974.

Smyth J.R. (1973): An orthopyroxene structure up to 850°C. Am. Mineral. 58, p. 636–648.

Stanek J. and Hafner S.S. (1988): Electric field gradient tensors of ^{57}Fe in orthorhombic (Mg,Fe)SiO_3. Hyperfine Int. 39, p. 253–267.

Steffen G., Langer K. and Seifert F. (1988): Polarized electronic absorption spectra of Synthetic (Mg,Fe)-orthopyroxenes, ferrosilite and Fe^{3+} - bearing ferrosilite. Phys. Chem. Minerals 16, p. 120–129.

Stöhr J. and Siegmann H. Magnetism: From Fundamentals to Nanoscale Dynamics. Springer Verlag Berlin Heidelberg, 2006.

Sueno S., Cameron M. and Prewitt C. (1976): Orthoferrosilite: High-temperature crystal chemistry. Am. Mineral. 61, p. 38–53.

Tanabe Y. and Sugano S. (1954): On the Absorption Spectra of Complex Ions I. J. Phys. Soc. Jpn. 9, p. 753–766.

Tanabe Y. and Sugano S. (1954): On the Absorption Spectra of Complex Ions II. J. Phys. Soc. Jpn. 9, p. 766–779.

Taran M.N., Langer K. (2001): Electronic absorption spectra of Fe^{2+} ions in oxygen-based rock-forming minerals at temperatures between 297 and 600 K. Phys. Chem. Minerals 28, p.199–210.

Taran M. and Langer K. (2003): Single-crystal high-pressure electronic absorption spectroscopy study of natural orthopyroxenes. Eur. J. Mineral. 15, p. 689–695.

Templeton D. Molecular and cellular iron transport. Marcel Dekker, Basel, 2002.

Terra J. and Guenzburger D. (1989): Electronic structure and isomer shifts of Sn halides. Phys. Rev. B. 39, p. 50–56.

Terra J., Guenzburger D. (1991): Isomer shifts and chemical bonding in crystalline Sn(II) and Sn(IV) compounds. J. Phys. Condens. Matter 3, p. 6763–6774.

Thomas L. (1927): The Calculation of atomic fields. Math. Proc. of Cambridge Philosophical Society. 23, p. 542–548.

Van Alboom A., De Grave E., Vandenberghe R.E. (1993): Study of the temperature dependence of the hyperfine parameters in two orthopyroxenes by [57]Fe Mössbauer spectroscopy. Phys. Chem. Minerals 20, p.263–275.

Van Alboom A., De Grave E., Vandenberghe R.E. (1994): Crystal-field interpretation of the temperature dependence of the [57]Fe Mössbauer quadrupole splitting in two orthopyroxenes. Hyperf. Inter. 91, p.703–707.

Van Vleck J. Theory of Electric and Magnetic Susceptibilities. Chapt. XII. Oxford Univ. Press, London and New York, 1932.

Vosko S., Wilk L. and Nusair M. (1980): Accurate spin–dependent electron liquid correlation energies for local spin density calculations: a critical analysis. Can. J. Phys. 58, p. 1200–1211.

Weber S.-U., Grodzicki M., Geiger C. A., Lottermoser W., Tippelt G., Redhammer G. J., Bernroider M., Amthauer G. (2007): ^{57}Fe Mössbauer measurements and electronic structure calculations on natural lawsonites. Phys. Chem. Minerals 34, p. 1–9.

Weber S.-U., Grodzicki M., Lottermoser W., Redhammer G. J., Tippelt G., Ponahlo J., Amthauer G. (2007): ^{57}Fe Mössbauer spectroscopy, X-ray single-crystal diffractometry, and electronic structure calculations on natural alexandrite. Phys. Chem. Minerals 34, p. 507–515.

Weber S.U., Grodzicki M., Lottermoser W., Redhammer G.J., Topa D., Tippelt G., Amthauer G. (2009): ^{57}Fe-Mössbauer spectroscopy, X-ray single-crystal diffractometry, and electronic structure calculations on natural sinhalites. Phys. Chem. Minerals 36, p. 259–269.

Weisskopf V. and Wigner E. (1930): Berechnung der natürlichen Linienbreite auf Grund der Diracschen Lichttheorie. Z. Phys., 63, p. 54–73.

Wildner M., Andrut M. and Rudowicz C. Optical absorption spectroscopy in geosciences. Part I: Basic concepts of crystal field theory. EMU notes in mineralogy. Eötvös University Press, Budapest, 2004, p. 93–144.

White W.B., Keester K.L. (1966): Optical absorption spectra of iron in the rock-forming silicates. Am. Mineral. 51, p. 774–791.

White B., Moore R. (1972): Interpretation of the spin-allowed bands of Fe^{2+} in silicate garnets. Am. Mineral. 57, p. 1692–1710.

Wiedenmann A., Regnard J.R. (1986): Neutron diffraction study of the magnetic ordering in pyroxenes $Fe_xMg_{1-x}SiO_3$. Sol. State. Commun. 57, p. 499–504.

Wiedenmann A., Regnard J.-R., Fillion G. and Hafner S.S. (1986): Magnetic properties and magnetic ordering of the orthopyroxenes $Fe_xMg_{1-x}SiO_3$. J.Phys. C: Solid State Phys. 19, p. 3683–3695.

Winkler W. Vetter R., Hartmann E. (1987): Mössbauer isomer shift calibration of ^{119}Sn using pseudopotential and Xα SW calculations. Chem. Phys. 114, p. 347–358.

Woodland A. and Ross C. (1994): A Crystallographic and Mössbauer Spectroscopy Study of $Fe^{2+}_3Al_2Si_3O_{12}$-$Fe^{2+}_3Fe^{3+}_2Si_3O_{12}$ (Almandine-"Skiagite") and $Ca_3Fe^{3+}_2Si_3O_{12}$-$Fe^{2+}_3Fe^{3+}_2Si_3O_{12}$ (Andradite-"Skiagite") Garnet Solid Solution. Phys. Chem. Minerals 21, p. 117–132.

Yanaga M., Endo K., Nakahara H., Ikuta S., Miura T., Takahashi M., Takeda M. (1990): Calibration of the isomer shift of ^{121}Sb and ^{119}Sb by means of Mössbauer spectroscopy and molecular orbital calculations. Hyperf. Interact. 62, p. 359–372.

Yang H. and Ghose S. (1995): A transitional structural state and anomalous Fe-Mg order-disorder in Mg-rich orthopyroxene, $(Mg_{0.75}Fe_{0.25})_2Si_2O_6$. Am. Mineral. 80, p. 9–20.

Zherebetskyy D., Amthauer G. and Grodzicki M. (2010): Electronic and magnetic structure of pyroxenes: II. Orthoferrosilite. Phys. Chem. Minerals, in Pub. DOI 10.1007/s00269-009-0346-7.

PUBLICATIONS AND PRESENTATIONS

Publications in Refereed Journals

Zherebetskyy D., Amthauer G. and Grodzicki M. (2010): Electronic and magnetic structure of pyroxenes: II. Orthoferrosilite. Phys. Chem. Minerals, in Pub. DOI 10.1007/s00269-009-0346-7.

Zherebetskyy D., Lebernegg S., Amthauer G., Grodzicki M. Electronic and magnetic structure of garnets: I. Almandine. Phys. Chem. Minerals, submitted.

Zherebetskyy D., Amthauer G., Grodzicki M. Electronic and magnetic structure of garnets: II. Andradite. Phys. Chem. Minerals, to be published.

Scharff P., Siegmund C., Risch K., Lysko I., Lysko O., Zherebetskyy D., Ivanisik A., Gorchinskiy A., Buzaneva E. (2005): Characterization of Water-Soluble Fullerene C_{60} Oxygen and Hydroxyl Group Derivatives for Photosensitizer. Fullerenes, Nanotubes and Carbon Nanostructures, vol.13, supplementary 1, p. 497-509.

Belousov I.V., Gorchinskiy A., Lytvyn P., Kuznetsov G., Popova G., Veblaya T., Zherebetskyy D., Lysko O., Vysokolyan O., Buzaneva E. (2003): Self formation of Si nanostructured layer at the metal silicide/silicon interface. Materials Science and Engineering C 23, p.181-186.

Buzaneva E.V., Karlash A., Yakovkin K., Shtogun Ya., Putselyk S., Zherebeskiy D., Gorchinskiy A., Popova G., Prilutska S., Matyshevska O., Prilutskyy Yu., Lytvyn P., Scharff P., Eklund P. (2002): DNA nanotechnology of carbon nanotube cells: physico-chemical models of self-organization and properties. Materials Science and Engineering C19, p.41-45.

Publications in Books

Khomenko A. and Zherebetskyy D. Characterization of nano- and microstructures from fullerene C_{60}/DNA on silicon substrate by optical, atomic force microscopy and electron, vibrational spectroscopy. In Book: Proceeding of the Fifth International Young Scientists' Conference on Applied Physics, June, 20-22, 2005, Kyiv, Ukraine, Taras Shevchenko National University of Kyiv, Faculty of RadioPhysics, p.162-163.

Buzaneva E., Gorchinskiy A., Scharff P., Risch K., Nassiopoulou A., Tsamis C., Prilutskyy Yu., Ivanyuta O., Zhugayevych A., Kolomiyets D., Veligura A., Lysko I., Vysokolyan O.,

Lysko O., Zherebetskyy D., Khomenko A., Sporysh I. DNA, DNA/metal nanoparticles, DNA/nanocarbon and macrocyclic metal complex/fullerene molecular building blocks for nanosystems: electronics and sensing. In Book Frontiers of multifunctional integrated nanosystems, Eds: Eugenia Buzaneva and Peter Scharff, NATO Science Series II, Mathematics, Physics and Chemistry – Vol. 152, Kluwer Academic Publishers, Dordrecht, 2004, p. 251-276.

Buzaneva E., Gorchinskiy A., Prilutskyy Yu., Ivanyuta O., Veligura A., Zherebetskyy D., Lysko O., Kolomiyets D., Vysokolyan O., Lysko I., Sporysh I., Bezugla O., Scharff P., Risch K., Tsamis C., Nassiopoulou A. Design and study of DNA/nanocarbon and macrocyclic metal complex/C_{60} nanostructures. In Work Book of NATO ARW/Summer School, Frontiers in molecular-scale science and technology of nanocarbon, nanosilicon and biopolymer integrated nanosystems, Ilmenau, Germany, July 12-16, 2003, p.21.

Lysko I., Lysko O., Zherebetskyy D. Optical and Electronic properties of Water-soluble fullerene C_{60} Oxygen and Hydroxyl group derivatives for photosensitizers // In Work Book of IV International Young Scientists Conference Problems of Optics and High Technology Material Science, Kiev, Ukraine, October 27-30, 2003, p.79.

Zherebetskyy D. The evidence of DNA and DNA:Au molecular networks organization by atomic force microscopy. In Work Book of NATO ARW/Summer School, Frontiers in molecular-scale science and technology of nanocarbon, nanosilicon and biopolymer integrated nanosystems, Ilmenau, Germany, 2003, p. 69.

Zherebetskyy D. Atomic force microscopy of biopolymers: DNA and DNA:Au molecular networks// In Proceedings Book III International Young Scientist Conference on Applied Physics, Kiev, Ukraine, June 18-20, 2003, p. 140.

Buzaneva E., Gorchynskyy A., Popova G., Karlash A., Shtogun Ya., Yakovkin K., Zherebetskiy D., Matyshevska O., Prylutskyy Yu., Scharff P. Nanotechlology of DNA/Nano-Si and DNA/Carbon Nanotubes/Nano-Si Chips, In Book, Frontiers of multifunctional nanosystems, Eds: Eugenia Buzaneva and Peter Scharff, NATO Science Series, II-Mathematics, Physics and Chemistry – Vol.57, Kluwer Academic Publishers, Dordrecht, 2002, p. 191-212.

Ivanyuta A., Vysokolyan O., Lysko O., Zherebeskyy D., Veligura A., Buzaneva E., Gorchinskiy A., Prilutskyy Yu., Scharff P., Risch K., Carta-Abelmann L. Recognition of DNA, DNA:Au, DNA:C_{60} clusters using optical spectroscopy and microwave resistance. In Book of III International Young Scientists Conference Problems of Optics and High Technology Material Science, 2002, Kiev (Ukraine), p. 14.

Ivanyuta A., Vysokolyan O., Lysko O., Zherebeskyy D., Veligura A., Buzaneva E., Gorchinskiy A ., Prilutskyy Yu., Scharff P., Risch K., Carta-Abelmann L. Nanostructured layers from DNA, DNA:Au, DNA:C_{60} clusters: optical properties and microwave resistance. In Work Book of NATO ARW Nanostructured Materials and Coatings for Biomedical and Sensor Applications, 2002, Kiev (Ukraine), PII.20.

Buzaneva E., Karlash A., Yakovkin K., Shtogun Y., Putselyk S., Zherebetskiy D., Gorchinskiy A., Popova G., Prilutska S., Matyshevska O., Prilutskyy Y., Lytvyn P., Sharff P., Eklund P. DNA Nanotechnology of Carbon Nanotube Cells: Physico-Chemical Models of Self-Organization and Properties. In Scientific Program Book E-MRS 2001, Strasbourg, Symposiums S, S/ PI.17.

Reports and Abstracts at International Meetings

Joint Annual Meeting of ÖPG/SPS/ÖGAA, 2-4 September 2009, Innsbruck, Austria "Magnetic structure of almandine: electronic structure calculations".

International Conference on Magnetism 2009 (ICM2009) July 26 - 31, 2009, Karlsruhe, Germany "Magnetic structure of Fe-bearing silicates: electronic structure calculations".

Seminar- Day Salzburg-München, July 07, 2009, "Magnetic structure of Fe-bearing silicates: electronic structure calculations".

86th Annual Meeting of the DMG (DMG 2008), Symposium 08: The Physics and Chemistry of Minerals, Berlin (Germany), 14-17 September 2008, "Electronic and magnetic structure of orthoferrosilite: Electronic structure calculations".

E-MRS 2006 Spring Meeting, Symposium M,M-11-05, Nice (France), May 29-June 2, 2006 "Nanophotonics of Donor – Acceptor pairs in conjugated polymer/C60(C70) oxygen derivative suspensions and films" (**awarded for Graduate Student Award**).

E-MRS 2005 Spring Meeting, Symposium F, F-PIII-35, Strasbourg (France), May 30-June 3, 2005 "Self-Assembled Thin Films from Building Blocks of Fullerene C_{60}/ C_{60} Oxygen Derivatives/ds-DNA on Silicon for Photovoltaic Chip: Nanostructure, Conductivity and Photosensitivity".

IV International Young Scientists Conference Problems of Optics and High Technology Material Science, Kiev (Ukraine), October 27-30, 2004, "Optical and Electronic properties of Water-soluble fullerene C_{60} Oxygen and Hydroxyl group derivatives for photosensitizers".

III International Young Scientist Conference on Applied Physics, Kiev (Ukraine), June 18-20, 2003.

NATO ARW / Summer School "Frontiers in molecular-scale science and technology of nanocarbon, nanosilicon and biopolymer integrated nanosystems", Ilmenau (Gernany), July 12-16, 2003.

E-MRS 2003 Sprint meeting, Symposium A. Current trends in nanoscience – from materials to application and Symposium B. Advanced multifunctional nanocarbon materials and nano systems, Strasbourg (France), June 10-13, 2003.

III International Young Scientists Conference Problems of Optics and High Technology Material Science, Kiev (Ukraine), October 24-26, 2002.

NATO ARW "Nanostructured Materials and Coatings for Biomedical and Sensor Applications", Kiev (Ukraine), August 1-5, 2002.

E-MRS 2002 Sprint meeting, Symposium Q, Current Trends in Nanotechnologies: From Materials to Systems, Strasbourg (France), June 18-21, 2002.

E-MRS 2001 Sprint meeting, Symposium S, Current Trends in Nanotechnologies: From Materials to Systems, Strasbourg (France), June 10-13, 2001.